PRAISE FOR MICHAEL BARTON

From "*The Power of Regenerative Agriculture:*"

This is your go-to book to help inspire and educate yourself about the possibility of how you can play a role in the interconnection of farming systems with the entirety of the ecological system.

— D. G. FARNSWORTH

Well done book. I liked the chapter summaries at the end of each chapter. Chapter 6, was spot on in describing the challenges and barriers to implementing regenerative agriculture. The author has a passion for this subject and has provided very clear and concise data. The Afterword ends the book on a positive note and a view of hope that we'll get there.

— SRP

These guides to being environmentally responsible often come across as a little preachy. This is not one of those. It's informative and well-written, with a cheerful attitude toward ecologically friendly farming methods.

— LWHIT

THE POWER OF GROWING TOGETHER

THE POWER OF GROWING TOGETHER

AN INTRODUCTION TO PRACTICAL PERMACULTURE, AGROFORESTRY, AND SILVOPASTURE FOR A REGENERATIVE AND SUSTAINABLE FUTURE (2-IN-1 COLLECTION)

SUSTAINABLE AGRICULTURE

MICHAEL BARTON

Book
Bound Studios

To Mother Earth and all the passionate gardeners, farmers, and dreamers who believe in a greener, more sustainable world.

To Mother Earth and all the passionate gardeners, farmers, and dreamers who believed in a more sustainable way.

In nature's infinite book of secrecy, a little I can read.

— WILLIAM SHAKESPEARE, ANTONY AND CLEOPATRA

CONTENTS

AGROFORESTRY AND SILVOPASTURE

PRACTICAL PERMACULTURE

A BEGINNER'S GUIDE TO DESIGNING A RESILIENT AND SUSTAINABLE GARDEN ECOSYSTEM FOR SELF-SUFFICIENT LIVING

INTRODUCTION

In today's fast-paced world, where convenience often takes precedence over sustainability, finding ways to live in harmony with our environment is more important than ever. One such approach is permaculture, a holistic design system that seeks to create sustainable human habitats by emulating the patterns and relationships found in nature. Practical permaculture is the application of these principles to our everyday lives, transforming our homes, gardens, and communities into thriving, resilient ecosystems.

The term "permaculture" is a combination of the words *"permanent" and "agriculture,"* reflecting the movement's origins in sustainable agriculture. However, permaculture has since evolved into various disciplines, including architecture, urban planning, and social systems. At its core, permaculture is about designing **systems that work with nature**, rather than against it, to meet our needs while preserving the planet's health.

In this chapter, we will introduce you to the world of practical permaculture, exploring its history, principles, and ethics. We will then delve into the key components of a permaculture system, from soil health and water management to plant selection and animal integration. Next, we will guide you through the process of designing your own permaculture garden, whether you have a small urban balcony or a sprawling rural property. Finally, we will share inspiring real-life examples and success stories from around the globe, demonstrating the transformative power of practical permaculture.

As you embark on this journey, you will discover that permaculture is more than just a set of gardening techniques; it is a mindset, a way of life that fosters a deep connection with the natural world and a sense of responsibility for its well-being. By embracing practical permaculture, you create a more sustainable future for yourself and your loved ones and contribute to the larger global movement towards a more just, equitable, and regenerative world. So let's get started, and welcome to the world of practical permaculture!

The Principles and Ethics of Permaculture

Permaculture is a holistic approach to sustainable living that integrates ecological design, resource conservation, and self-sufficiency. At the heart of permaculture are its principles and ethics, which serve as guidelines for creating harmonious, resilient, and productive systems that benefit both humans and the environment. This section will delve into these foundational concepts and explore how they can be applied practically to create thriving permaculture systems.

The Ethics of Permaculture

Permaculture is grounded in three core ethics that guide its practitioners in making responsible and sustainable choices. These ethics are:

Earth Care: This ethic emphasizes the importance of nurturing and preserving the planet's ecosystems, soil, water, and biodiversity. Earth care encourages us to consider the long-term consequences of our actions and to strive for practices that regenerate and enhance the environment.

People Care: This ethic fosters healthy, supportive, and resilient communities. People care encourages cooperation, sharing of resources, and prioritizing the well-being of all individuals. It also highlights the importance of personal growth, self-reliance, and empowerment.

Fair Share: This ethic promotes equitable distribution of resources and opportunities within human societies and between humans and other species. Fair sharing encourages us to limit our consumption, reduce waste, and share surplus resources with others in need.

The Principles of Permaculture

Permaculture principles, derived from observing natural ecosystems, provide practical guidance for designing and managing sustainable systems. While there are various interpretations of these principles, the following twelve are widely recognized and used by permaculture practitioners:

1. **Observe and Interact:** Careful observation of natural systems and thoughtful interaction with them allows us to learn from nature and adapt our practices accordingly.
2. **Catch and Store Energy:** Harnessing and storing renewable energy sources, such as sunlight, wind, and water, ensures a consistent supply of resources for our needs.
3. **Obtain a Yield:** Design systems that provide diverse and abundant resources, including food, energy, and materials, to meet our needs and support our well-being.
4. **Apply Self-Regulation and Accept Feedback:** Recognize the limits of our actions and adapt our practices based on the feedback we receive from our systems and the environment.
5. **Use and Value Renewable Resources and Services:** Prioritize using renewable resources and natural processes to minimize our dependence on non-renewable inputs.
6. **Produce No Waste:** Minimize waste by valuing and using all resources, recycling, and repurposing materials whenever possible.

7. **Design from Patterns to Details:** Observe patterns in nature and use them as a foundation for designing systems while also considering the unique details of each situation.
8. **Integrate Rather Than Segregate:** Create synergistic relationships between elements in a system, promoting cooperation, resource sharing, and mutual support.
9. **Use Small and Slow Solutions:** Emphasize small-scale, localized solutions that are more manageable, adaptable, and resilient over time.
10. **Use and Value Diversity:** Incorporate diverse elements, species, and strategies to create resilient systems that can withstand disturbances and adapt to change.
11. **Use Edges and Value the Marginal:** Recognize the potential in transitional areas and underutilized spaces where unique interactions and opportunities can occur.
12. **Creatively Use and Respond to Change:** Embrace change as an opportunity for growth and adaptation, and design systems that can evolve over time.

By understanding and applying these ethics and principles, we can create permaculture systems that are sustainable and regenerative, improving the health of our planet and our communities. In the following sections, we will explore the key components of a permaculture system and learn how to design and implement practical permaculture solutions in our own gardens and landscapes.

Key Components of a Permaculture System

Permaculture is a holistic approach to gardening and agriculture that seeks to create sustainable, self-sufficient ecosystems. To achieve this, permaculture systems incorporate a variety of key components that work together to create a harmonious and productive environment. In this section, we will explore some of the most important elements of a permaculture system and how they contribute to its overall success.

Zones and Sectors

One of the foundational concepts in permaculture is the idea of dividing your land into zones and sectors. Zones are garden or landscape areas organized by the frequency of human use and interaction. For example, Zone 1 might include the areas closest to your home, such as a kitchen garden or herb spiral, while Zone 5 might be a more remote area designated for wildlife habitat and minimal human intervention.

Sectors, on the other hand, are determined by the natural elements that influence your site, such as sunlight, wind, and water. By understanding these factors and designing your garden accordingly, you can maximize the efficiency and productivity of your permaculture system.

Plant Stacking and Polycultures

In a permaculture system, plants are strategically placed to maximize their benefits and minimize competition for resources. This is achieved through a technique called plant stacking, which involves layering plants vertically to create a diverse and productive ecosystem. For example, tall trees provide shade for understory plants, while groundcovers and root crops fill in the spaces below.

Polycultures, or the practice of growing multiple species together, also play a crucial role in permaculture systems. By planting a variety of crops, you can create a more resilient and productive garden that is less susceptible to pests and diseases.

Soil Health and Fertility

Healthy soil is the foundation of any successful permaculture system. You can support a diverse and productive ecosystem by focusing on building and maintaining soil fertility. Permaculture techniques for improving soil health include composting, mulching, and incorporating nitrogen-fixing plants into your garden design.

Water Management

Water is a precious resource, and permaculture systems aim to use it as efficiently as possible. This can be achieved through a variety of methods, such as capturing and storing rainwater, creating swales to slow and spread water across the landscape, and using drip irrigation systems to minimize water waste.

Animal Integration

Animals play a vital role in permaculture systems, providing valuable services such as pest control, soil aeration, and nutrient cycling. You can create a more balanced and productive ecosystem by incorporating animals into your garden design. Examples of animal integration include keeping chickens for pest control and egg production or incorporating a pond for fish and aquatic plants.

Energy Efficiency and Renewable Resources

Permaculture systems strive to minimize their reliance on non-renewable resources and instead focus on using energy and materials as efficiently as possible. This might include using solar panels for electricity, incorporating passive solar design principles into your home, or utilizing locally sourced and recycled materials for construction projects.

By understanding and incorporating these key components into your permaculture system, you can create a thriving and sustainable ecosystem that provides for your needs while benefiting the environment. As you continue to learn and experiment with permaculture techniques, you'll discover the incredible potential of this holistic approach to gardening and agriculture.

Designing Your Permaculture Garden

Embarking on the journey to create your own permaculture garden is exciting and rewarding. This section will guide you through the process of designing a garden that is both practical and sustainable while also being a beautiful and productive space. By following the principles and ethics of permaculture, you can create a garden that works in harmony with nature, providing you with an abundance of fresh produce and a thriving ecosystem.

Assessing Your Site

Before diving into the design process, it's essential to assess the unique characteristics of your site. This includes climate, topography, soil type, and available resources. By understanding the strengths and limitations of your site, you can make informed decisions about which plants and techniques will work best in your permaculture garden.

Begin by observing your site throughout the day and across different seasons. Take note of sun exposure, wind patterns, and water flow. This will help you determine the ideal placement for various garden elements, such as planting beds, water features, and windbreaks.

Zoning and Sector Analysis

In permaculture design, zoning is a technique used to organize your garden based on the frequency of human use and plant or animal needs. The zones are numbered from 1 to 5, with Zone 1 being the area closest to your home and Zone 5 being the furthest away.

- **Zone 1:** This area is visited daily and should include elements that require frequent attention, such as herbs, salad greens, and small fruit trees.
- **Zone 2:** This zone is visited less frequently and is suitable for larger fruit trees, vegetables, and small livestock.

- **Zone 3:** This area is used for larger-scale crop production and may include orchards, pasture, or grain fields.
- **Zone 4:** This zone is semi-wild and can be used for foraging, firewood production, and wildlife habitat.
- **Zone 5:** This area is left undisturbed as a natural ecosystem, providing a space for wildlife and natural processes to occur.

Sector analysis involves identifying external factors that may influence your garden, such as sun, wind, and water. By understanding these factors, you can design your garden to maximize positive influences and minimize negative ones.

Choosing Plants and Animals

When selecting plants and animals for your permaculture garden, consider their functions, needs, and relationships. Choose plants that serve multiple purposes, such as providing food, improving soil fertility, and attracting beneficial insects. Consider incorporating animals like chickens or bees to help with pest control, pollination, and nutrient cycling.

Creating a Garden Layout

With your site assessment, zoning, and plant selection complete, it's time to create a layout for your permaculture garden. Start by sketching a rough plan, placing elements based on their zone and sector requirements. Consider the relationships between plants and animals, ensuring each element supports and benefits from its neighbors.

Incorporate techniques such as swales, hugelkultur beds, and keyhole gardens to maximize water efficiency and soil fertility. Design pathways and access points to make it easy to maintain and harvest your garden.

Implementing and Adapting Your Design

Once your design is complete, it's time to bring your permaculture garden to life. Start by preparing the soil, adding organic matter, and creating planting beds. Next, plant your chosen species, taking care to provide them with the appropriate conditions for growth.

As your garden matures, observe and learn from its successes and challenges. Adapt your design as needed, making adjustments to improve productivity and sustainability. Remember that permaculture is an ongoing process, and your garden will evolve over time.

By following these steps, you can create a permaculture garden that is both practical and sustainable, providing you with an abundance of fresh produce and a thriving ecosystem. Embrace permaculture principles and enjoy the journey towards a more sustainable future.

Real-Life Examples and Success Stories

This section will explore some inspiring real-life examples and success stories of practical permaculture. These stories demonstrate the transformative power of permaculture principles when applied to various contexts, from urban gardens to rural farms. By examining these examples, we can gain valuable insights into the potential of permaculture to create sustainable, productive, and resilient systems that benefit both people and the planet.

The Urban Oasis: A Permaculture Garden in the City

Our first example takes us to the heart of the bustling city in *San Francisco*, where a small plot of land has been transformed into a thriving permaculture garden. This urban oasis demonstrates that even in the most densely populated areas, it is possible to create productive and sustainable green spaces.

The garden's design maximizes the use of vertical space by implementing trellises, arbors, and stacked planters. This allows for a diverse range of plants to be grown, including vegetables, herbs, and fruit trees. **Rainwater harvesting** systems and composting methods are employed to minimize waste and promote a closed-loop system. The garden also serves as a community hub where neighbors can come together to learn about permaculture, share resources, and enjoy the fruits of their labor.

The Regenerative Farm: Healing the Land and Building Community

Next, we visit a rural farm in *Missouri* that has embraced permaculture principles to regenerate degraded land and build a strong, resilient community. This farm's success story demonstrates the power of permaculture to heal ecosystems and foster social connections.

The farm has dramatically increased soil fertility, biodiversity, and overall productivity by implementing agroforestry, holistic grazing, and no-till farming practices. The farm also hosts workshops and educational programs, teaching others about the benefits of regenerative agriculture and permaculture design. By fostering a sense of community and shared responsibility, this farm has become a model for sustainable agriculture and rural development.

The Schoolyard Transformation: Educating the Next Generation

Our final example highlights the transformative power of permaculture in an educational setting. A once-barren schoolyard in *California* has been transformed into a vibrant permaculture garden, providing students with hands-on learning opportunities and a deeper connection to the natural world.

The garden incorporates a variety of permaculture design elements, including raised beds, companion planting, and a small-scale aquaponics system. Students are actively involved in the garden's maintenance and harvest, learning valuable skills in organic gardening, composting, and water conservation. The garden also serves as an outdoor classroom, where lessons in ecology, biology, and **environmental stewardship** come to life.

These real-life examples and success stories demonstrate the incredible potential of practical permaculture to create positive change in diverse settings. By applying the principles and ethics of permaculture, we can design systems that not only provide for our needs but also regenerate ecosystems, build community, and foster a more sustainable future for all.

Embracing a Sustainable Future with Practical Permaculture

As we reach the end of this brief introduction to the world of practical permaculture, it is important to reflect on the powerful impact that this holistic approach to gardening and land management can have on our lives, our communities, and our planet. By embracing the principles and ethics of permaculture, we can create sustainable, resilient, and abundant systems that not only provide for our needs but also contribute to the healing and regeneration of our environment.

To conclude, let us remember the words of permaculture co-founder *Bill Mollison: "Though the problems of the world are increasingly complex, the solutions remain embarrassingly simple."* By applying the wisdom of practical permaculture, we can be part of the solution and help create a world where all living beings can thrive.

1

UNDERSTANDING PERMACULTURE PRINCIPLES

A flourishing permaculture garden with diverse plants and wildlife.

Permaculture is a holistic approach to designing sustainable systems that work in harmony with nature. It is a practical and ethical framework that aims to create **resilient, self-sufficient,** and **regenerative environments.** In this chapter, we will delve into the core principles of permaculture, which serve as a foundation for designing and implementing sustainable systems in various aspects of our lives.

The concept of permaculture was developed in the *1970s* by Australian ecologists Bill Mollison and David Holmgren. They sought to create a solution to the growing environmental and social challenges of their time, such as soil erosion, deforestation, and resource depletion. Permaculture is not just about growing food; it is a comprehensive approach encompassing various aspects of human life, including housing, energy, water management, and community development.

At the heart of permaculture lie its principles, which provide a set of guidelines for designing sustainable systems. These principles are derived from observing and understanding the patterns and relationships found in nature. By mimicking these natural processes, we can create systems that are not only sustainable but also regenerative, meaning they improve the environment and the quality of life for all living beings.

In this chapter, we will explore the ethics and principles of permaculture, which serve as a foundation for its practice. First, we will discuss the three core ethics of permaculture: Earth Care, People Care, and Fair Share, which guide our actions and decisions in creating sustainable systems. We will then delve into the twelve design principles of permaculture, which provide a framework for designing and implementing these systems in various aspects of our lives.

Furthermore, we will discuss how to apply permaculture principles in everyday life, from our homes and gardens to our communities and workplaces. We will also explore real-world examples of permaculture in action, showcasing the incredible potential of this approach to create positive change in the world.

By understanding and embracing permaculture principles, we can work towards a more sustainable, resilient, and regenerative future for ourselves, our communities, and the planet as a whole. So, let's embark on this journey together and discover the power of permaculture to transform our lives and the world around us.

The Ethics of Permaculture: Earth Care, People Care, and Fair Share

At the heart of permaculture lies a set of three core ethics that serve as guiding principles for designing sustainable systems. These ethics - Earth Care, People Care, and Fair Share - are not only essential for the success of permaculture projects but also for fostering a sense of responsibility and interconnectedness among individuals and communities. In this section, we will delve into each of these ethics and explore their significance in the realm of permaculture.

Earth Care: Nurturing Our Planet

The first ethic, Earth Care, emphasizes the importance of protecting and preserving our planet's ecosystems, natural resources, and biodiversity. This ethic encourages us to adopt practices that regenerate and enhance the environment rather than deplete or harm it. By nurturing the earth, we ensure it remains a thriving and bountiful home for all living beings.

Earth Care is practiced in permaculture by designing systems that work in harmony with nature rather than against it. This includes using renewable resources, minimizing waste, and promoting biodiversity. For example, a permaculture garden may incorporate plants that attract beneficial insects, create habitat for wildlife, and improve soil fertility, all while providing food and other resources for humans.

People Care: Supporting Our Communities

The second ethic, People Care, focuses on the well-being of individuals and communities. Permaculture recognizes that the health of our planet is inextricably linked to the health of its inhabitants. By fostering solid and resilient communities, we create a support network that can better withstand the challenges of a changing world.

People Care is practiced in permaculture by designing systems that meet the needs of all community members, including access to healthy food, clean water, and safe shelter. Additionally, permaculture encourages sharing knowledge and skills, empowering individuals to become active participants in their own well-being and that of their community. This can be achieved through workshops, skill-sharing events, and community projects that bring people together to learn and grow.

Fair Share: Distributing Resources Equitably

The third ethic, Fair Share, is centered on the equitable distribution of resources and opportunities. This ethic acknowledges that our planet's resources are finite, and we must use them wisely to ensure that all living beings can thrive. By embracing the concept of "**enough,**" we can create systems that provide for everyone without exceeding the earth's carrying capacity.

In permaculture, Fair Share is practiced by designing systems that minimize consumption and waste while maximizing the use of available resources. This may involve using energy-efficient technologies, repurposing materials, and sharing surplus resources with others. For example, a permaculture community might establish a tool-sharing program, allowing members to access tools and equipment without the need for each individual to purchase their own.

By understanding and embracing the ethics of Earth Care, People Care, and Fair Share, we can create permaculture systems that provide for our needs and contribute to the well-being of our planet and its inhabitants. In addition, these ethics serve as a foundation for sustainable living, inspiring us to cultivate a deeper connection with the earth and each other.

The Twelve Design Principles of Permaculture

Permaculture is a holistic approach to sustainable living, integrating various elements of ecology, agriculture, and community design. At the heart of permaculture are twelve design principles that serve as a guide for creating harmonious, resilient, and productive systems. These principles, developed by permaculture co-founder David Holmgren, can be applied to any scale, from a small backyard garden to an entire community or ecosystem. Let's explore each of these principles and understand their significance in permaculture design.

1. **Observe and Interact:** The first principle emphasizes the importance of observation and engagement with the natural environment. By carefully observing and understanding the patterns, relationships, and interactions within a system, we can make informed decisions and design solutions that are in harmony with nature.

2. **Catch and Store Energy:** This principle encourages us to capture and store energy in various forms, such as water, sunlight, and biomass. By doing so, we can create a reserve of resources that can be utilized during times of scarcity or need.

3. **Obtain a Yield:** Permaculture systems should be designed to provide for our needs, whether food, shelter, or other resources. We can create a sustainable and self-reliant system by ensuring we obtain a yield from our efforts.

4. **Apply Self-Regulation and Accept Feedback:** This principle highlights the importance of creating systems that are self-regulating and responsive to feedback. By designing systems that can adapt and evolve, we can minimize the need for external inputs and reduce the risk of system failure.

5. **Use and Value Renewable Resources and Services:** Permaculture encourages the use of renewable resources and natural services, such as solar energy, wind power, and natural pest control. Utilizing these resources can reduce our dependence on non-renewable resources and minimize our ecological footprint.

6. **Produce No Waste:** This principle emphasizes the importance of minimizing waste and utilizing resources efficiently. By designing

systems that recycle and reuse materials, we can create a closed-loop system that mimics the efficiency of natural ecosystems.

7. Design from Patterns to Details: Permaculture design starts with understanding the larger patterns and relationships within a system and then moves toward the specific details. By focusing on the broader context, we can create more resilient and adaptable systems.

8. Integrate Rather Than Segregate: In permaculture, elements within a system are designed to work together and support one another. We can create a more efficient and harmonious system by creating connections and fostering relationships between different components.

9. Use Small and Slow Solutions: This principle encourages us to focus on small-scale, manageable solutions that can be implemented gradually over time. By starting small and building upon our successes, we can create lasting change and minimize the risk of failure.

10. Use and Value Diversity: Diversity is a key component of resilient systems, as it provides redundancy and increases the overall stability of a system. By incorporating a variety of plants, animals, and strategies, we can create a more adaptable and resilient system.

11. Use Edges and Value the Marginal: In permaculture, the edges and marginal areas of a system are often the most productive and diverse. By focusing on these areas and utilizing their unique characteristics, we can create innovative solutions and maximize the potential of our systems.

12. Creatively Use and Respond to Change: Change is an inevitable part of life, and permaculture encourages us to embrace and adapt to these changes. By being flexible and responsive, we can turn challenges into opportunities and create systems that are resilient in the face of uncertainty.

By understanding and applying these twelve design principles, we can create permaculture systems that are not only sustainable but also regenerative, improving the health of our environment and the well-being of our communities. The following sections will explore how these principles can be applied in everyday life and examine real-world examples of permaculture in action.

Applying Permaculture Principles in Everyday Life

Incorporating permaculture principles into your everyday life may seem daunting, but it doesn't have to be. By making small, conscious changes and adopting a permaculture mindset, you can create a more sustainable and harmonious lifestyle. This section will explore practical ways to apply permaculture principles in your daily life, making it easier for you to embrace these concepts and contribute to a more sustainable future.

To begin with, let's consider the three ethics of permaculture: Earth Care, People Care, and Fair Share. These ethics can guide your actions and decisions in various aspects of your life, from the food you eat to the way you interact with your community. Here are some suggestions for incorporating these ethics into your daily routine:

Earth Care: Make an effort to reduce your ecological footprint by conserving resources, reducing waste, and supporting local, organic, and sustainable products. This can include simple actions like turning off lights when not in use, recycling, and composting your food waste. Additionally, consider growing your own food or supporting local farmers' markets to promote sustainable agriculture.

People Care: Foster a sense of community and connection by participating in local events, volunteering, or joining a community garden. You can create a supportive network that values collaboration and mutual aid by nurturing relationships with those around you.

Fair Share: Practice mindful consumption by being aware of the impact your choices have on the environment and other people. This can involve reducing your consumption of non-essential items, sharing resources with others, and supporting businesses that prioritize ethical and sustainable practices.

Next, let's explore how you can apply the twelve design principles of permaculture in your everyday life:

1. Observe and interact: Pay attention to the natural patterns and cycles around you, such as the changing seasons, local wildlife, and weather patterns. This awareness can help you make informed decisions about your lifestyle and surroundings.

2. Catch and store energy: Utilize renewable energy sources, such as solar panels or wind turbines, to power your home. Additionally, consider insulating your home to conserve energy and reduce heating and cooling costs.

3. Obtain a yield: Grow your own food or support local agriculture to ensure a sustainable and nutritious food supply.

4. Apply self-regulation and accept feedback: Continuously evaluate and adjust your actions and choices based on their impact on the environment and your community.

5. Use and value renewable resources and services: Prioritize the use of renewable resources, such as solar power, rainwater harvesting, and sustainable materials in your home and garden.

6. Produce no waste: Adopt a zero-waste mindset by reducing, reusing, and recycling materials whenever possible.

7. Design from patterns to details: Observe the patterns in nature and apply them to your home, garden, and community to create a harmonious and efficient environment.

8. Integrate rather than segregate: Foster connections between different elements of your life, such as incorporating edible plants into your landscaping or inviting neighbors to participate in a community garden.

9. Use small and slow solutions: Focus on making incremental changes and adopting sustainable practices over time rather than seeking immediate and drastic transformations.

10. Use and value diversity: Embrace the unique qualities of your local environment, culture, and community by incorporating diverse plants, animals, and traditions into your lifestyle.

11. Use edges and value the marginal: Recognize the potential of underutilized spaces, such as vacant lots or neglected areas in your community, and transform them into productive and vibrant spaces.

12. Creatively use and respond to change: Embrace change as an opportunity for growth and adaptation, and be open to new ideas and perspectives.

Integrating these permaculture principles into your daily life can create a more sustainable, resilient, and fulfilling lifestyle. As you continue to learn and grow, you'll find that the benefits of permaculture extend far beyond your own backyard, contributing to a healthier and more harmonious world for all.

Permaculture in Action: Real-World Examples

As we delve deeper into the world of permaculture, it's essential to see how these principles can be applied in real-life situations. By examining various examples from around the globe, we can better understand how permaculture practices can be adapted to different environments and cultures. In this section, we will explore a few inspiring real-world examples that showcase the power of permaculture in action.

The Greening the Desert Project, Jordan

The Greening the Desert Project, located in the arid region of Jordan, is a testament to the transformative power of permaculture. This project, led by permaculture expert Geoff Lawton, has successfully turned a barren landscape into a thriving food forest. By implementing permaculture principles such as water harvesting, soil regeneration, and planting a diverse range of edible plants, the project has created a sustainable ecosystem that provides food, shelter, and income for the local community.

The Bullock Brothers Homestead, USA

The Bullock Brothers Homestead, located on Orcas Island in Washington State, is a prime example of a successful permaculture farm. For over 30 years, the Bullock brothers have been applying permaculture principles to their 25-acre property, resulting in a lush, productive landscape. The farm features a diverse array of plants, animals, and natural building techniques, all working together to create a self-sustaining ecosystem. In addition, the Bullock Brothers Homestead serves as an educational center, inspiring visitors to embrace permaculture principles in their own lives.

The Chikukwa Project, Zimbabwe

In the remote region of Chikukwa, Zimbabwe, a community-led permaculture project has transformed the lives of over 7,000 people. Faced with deforestation, soil erosion, and food scarcity, the Chikukwa community adopted permaculture principles to restore their land and improve their livelihoods. Through a combination of agroforestry, water management, and organic farming techniques, the Chikukwa Project has increased crop yields, improved food security, and empowered the community to become more self-reliant.

Incredible Edible, UK

Incredible Edible is a grassroots movement that began in the small town of Todmorden, UK. The initiative encourages residents to grow food in public spaces like parks, sidewalks, and even outside the police station. By embracing the permaculture principle of "people care," Incredible Edible has fostered a strong sense of community and connection to the local environment. The movement has since spread to over 100 communities worldwide, demonstrating the power of permaculture to inspire positive change on a global scale.

These real-world examples illustrate the incredible potential of permaculture to transform our landscapes, communities, and lives. By applying the principles of permaculture in our gardens, homes, and neighborhoods, we can contribute to a more sustainable and resilient future for our planet.

Embracing Permaculture for a Sustainable Future

As we reach the end of this chapter, it is essential to reflect on the significance of these principles and their potential to reshape our future. The beauty of permaculture lies in its simplicity, adaptability, and the fact that it is rooted in the natural world. By embracing these principles, we can create a more sustainable, resilient, and harmonious environment for ourselves and future generations.

The ethics of permaculture - Earth Care, People Care, and Fair Share - serve as a guiding compass, reminding us of our responsibility to protect the planet, nurture our communities, and promote equitable distribution of resources. These ethics are not just applicable to gardening or agriculture; they can be integrated into every aspect of our lives, from our homes and workplaces to our relationships and personal choices.

The Twelve Design Principles of Permaculture provide a practical framework for implementing these ethics in our daily lives. We can create efficient, productive, and self-sustaining systems by observing and interacting with our environment, catching and storing energy, obtaining a yield, and applying self-regulation and feedback. Furthermore, by using and valuing renewable resources, producing no waste, and designing from patterns to details, we can minimize our ecological footprint and maximize our positive impact on the planet.

Applying permaculture principles in everyday life may seem daunting at first, but it is essential to remember that small, incremental changes can lead to significant transformations. Start by making conscious choices in your home, such as reducing waste, conserving water, and growing your own food. Then, as you gain confidence and experience, you can expand your permaculture practices to your community, workplace, and beyond.

In conclusion, embracing permaculture for a sustainable future is a viable and necessary option. As we face unprecedented environmental challenges, it is crucial that we adopt practices that promote ecological balance, social equity, and economic stability. By incorporating permaculture principles into our lives, we can become active participants in creating a more sustainable, resilient, and harmonious world. So, let us embark on this journey together, guided by the wisdom of nature and the power of permaculture, and leave a lasting legacy for generations to come.

Chapter Summary

- Permaculture is a holistic approach to designing sustainable systems that work in harmony with nature, encompassing various aspects of human life, including housing, energy, water management, and community development.
- The three core ethics of permaculture are Earth Care, People Care, and Fair Share, which guide our actions and decisions in creating sustainable systems.
- The Twelve Design Principles of Permaculture provide a practical framework for implementing sustainable practices in various aspects of our lives.
- Permaculture principles can be applied in everyday life through small, conscious changes, such as reducing waste, conserving water, and growing your own food.
- The ethics and principles of permaculture can be integrated into every aspect of our lives, from our homes and workplaces to our relationships and personal choices.
- Real-world examples of permaculture in action demonstrate its incredible versatility and effectiveness in diverse settings across the globe.
- Embracing permaculture for a sustainable future is not only a viable option but a necessary one as we face unprecedented environmental challenges.
- By incorporating permaculture principles into our lives, we can become active participants in creating a more sustainable, resilient, and harmonious world for future generations.

2

DESIGNING YOUR PERMACULTURE GARDEN

A permaculture garden brimming with life, where diverse plants and vegetables grow harmoniously alongside each other, creating a lush green tapestry.

Welcome to the exciting journey of designing your very own permaculture garden! This chapter aims to guide you through the essential steps of creating a sustainable, productive, and beautiful space that works in harmony with nature. By embracing the principles of permaculture, you will not only enhance your garden's resilience and productivity but also contribute to a healthier planet.

Designing a permaculture garden is a creative and rewarding process that requires careful observation, planning, and implementation. The key to success lies in understanding the unique characteristics of your site and selecting the right combination of plants, animals, and sustainable techniques that work together to create a thriving ecosystem.

In this chapter, we will explore the following essential steps in designing your permaculture garden:

1. **Assessing Your Site and Its Resources:** Before you start designing your garden, it's crucial to understand the unique features and resources of your site. This section will guide you through the process of analyzing your soil, climate, water availability, and other factors that will influence your garden's design.

2. **Choosing the Right Plants and Animals for Your Garden:** Selecting the right combination of plants and animals is at the heart of permaculture design. This section will help you choose species that are well-suited to your site's conditions and that support each other in a mutually beneficial way.

3. **Integrating Sustainable Techniques and Systems:** Permaculture gardens are designed to be low-maintenance and self-sufficient. This section will introduce you to a range of sustainable techniques and systems that can help you conserve water, build soil fertility, and reduce waste in your garden.

4. **Creating a Maintenance Plan for Long-Term Success:** A well-designed permaculture garden should require minimal intervention to thrive. This section will provide guidance on creating a maintenance plan that ensures the long-term health and productivity of your garden.

5. Bringing Your Permaculture Garden to Life: Finally, we will wrap up the chapter with some tips and inspiration for transforming your garden design into a living, breathing permaculture paradise.

By following these steps and embracing the principles of permaculture, you will be well on your way to creating a garden that is not only beautiful and productive but also a shining example of sustainability and harmony with nature. So, let's dive in and start designing your permaculture garden!

Assessing Your Site and Its Resources

Before diving into the exciting world of permaculture design, it is crucial to take a step back and **assess the unique characteristics of your site**. This will help you make informed decisions and ensure your permaculture garden is tailored to your specific environment. This section will discuss the key aspects to consider when evaluating your site and its resources.

First and foremost, take note of the size and shape of your garden. This will determine the scale of your permaculture project and influence the layout of your design. Next, sketch a rough map of your site, marking any existing structures, pathways, and vegetation. This will serve as a helpful reference throughout the design process.

Next, observe the natural elements of your site, such as sunlight, wind, and water. Identify the sunniest and shadiest areas and the direction of prevailing winds. This information will guide your plant and animal selections and the placement of key elements like windbreaks and water catchment systems.

Soil quality is another crucial factor to consider. Conduct a simple soil test to determine the texture, pH, and nutrient levels of your garden's soil. This will help you identify any amendments needed to create a fertile and productive growing environment. Additionally, observe any existing drainage patterns and consider how they may impact your design.

Take a moment to appreciate the biodiversity of your site. Are there native plants, insects, or animals that already call your garden home?

These species can be valuable allies in your permaculture journey, contributing to pest control, pollination, and soil health. Make a list of these existing resources and consider how they can be integrated into your design.

Finally, assess the human and material resources available to you. Consider your skills, knowledge, time, and the support of friends, family, or neighbors who may be interested in collaborating on your permaculture project. Additionally, take stock of any materials you already have on hand, such as compost, mulch, or tools. These resources can reduce the overall cost of your project and contribute to its sustainability.

In conclusion, assessing your site and its resources is vital in designing a thriving permaculture garden. By understanding the unique characteristics of your environment and harnessing the power of available resources, you can create a thriving ecosystem that is both productive and sustainable.

Choosing the Right Plants and Animals for Your Garden

Selecting the appropriate plants and animals for your permaculture garden is a crucial step in ensuring its success. By carefully considering the needs and characteristics of each species, you can create a thriving ecosystem that benefits both you and the environment. In this section, we will explore the factors to consider when choosing plants and animals for your garden, as well as some popular choices for permaculture enthusiasts.

When selecting plants for your permaculture garden, it's essential to consider the following factors:

Climate and microclimate: Choose plants that are well-suited to your region's climate and the specific microclimate of your garden. This will ensure that your plants can thrive without excessive intervention or the need for artificial heating or cooling.

Soil type and fertility: Different plants have varying preferences for soil type and fertility levels. By selecting plants that are well-suited to your garden's soil conditions, you can minimize the need for soil amendments and fertilizers.

Sunlight and shade: Consider the amount of sunlight and shade available in your garden when selecting plants. Some plants require full sun, while others prefer partial shade or even full shade.

Water requirements: Choose plants with water requirements that match the natural rainfall patterns in your area, as well as the water resources available in your garden. This will help to minimize the need for supplemental irrigation.

Plant relationships: In a permaculture garden, plants are chosen for their individual characteristics and ability to interact positively with other plants and animals. Consider selecting plants that can provide multiple functions, such as nitrogen-fixing legumes, insect-attracting flowers, or plants that provide food and habitat for beneficial animals.

Some popular plant choices for permaculture gardens include fruit and nut trees, berry bushes, perennial vegetables, herbs, and edible flowers. These plants can provide food, medicine, and other resources while also contributing to a healthy and diverse ecosystem.

In addition to plants, incorporating animals into your permaculture garden can provide numerous benefits, such as pest control, soil fertility, and food production. When selecting animals for your garden, consider the following factors:

Size and space requirements: Choose animals that are appropriate for the size of your garden and the available space. For example, small gardens may be better suited to chickens or rabbits, while larger gardens may be able to accommodate goats or sheep.

Dietary needs: Consider the dietary needs of the animals you choose and ensure that your garden can provide adequate food resources. For example, chickens can forage for insects and plants, while goats and sheep require access to pasture or browse.

Compatibility with plants and other animals: Select animals that can coexist peacefully with your plants and other animals in the garden. For example, ducks can be an excellent addition to a garden with a pond, as they can help control pests and algae while also providing eggs and meat.

Local regulations and restrictions: Be sure to research any regulations or restrictions on keeping animals in your area before adding them to your permaculture garden.

Some popular animal choices for permaculture gardens include chickens, ducks, bees, rabbits, and even small livestock like goats and sheep. By carefully selecting the right plants and animals for your garden, you can create a thriving, sustainable ecosystem that provides abundant resources for you and your family.

Integrating Sustainable Techniques and Systems

Creating a thriving permaculture garden involves more than just selecting the right plants and animals. It requires integrating sustainable techniques and systems that work in harmony with nature, ensuring your garden's long-term health and productivity. This section will explore some of the most effective and widely-used sustainable practices that can be easily incorporated into your permaculture design.

Water Conservation and Management: One of the core principles of permaculture is to use resources efficiently, and water is no exception. You can significantly reduce your garden's reliance on external water sources by implementing rainwater harvesting, swales, and mulching strategies. Additionally, consider incorporating drought-tolerant plants and using greywater recycling systems to further minimize water waste.

Soil Health and Fertility: Healthy soil is the foundation of a thriving permaculture garden. To maintain and improve soil fertility, incorporate organic matter through composting, cover cropping, and the use of green manures. These practices add essential nutrients to the soil and improve its structure and water retention capacity. Furthermore, avoid tilling the soil, as this can disrupt the natural balance of microorganisms and lead to soil compaction.

Policulture and Plant Diversity: Planting diverse species in your garden promotes a more resilient and productive ecosystem. Polyculture, or growing multiple species together, can help reduce pests and diseases, improve soil fertility, and create a more visually appealing landscape. When selecting plants for your garden, consider their various functions, such as nitrogen-fixing, pollinator-attracting, or pest-repelling properties.

Natural Pest Management: Chemical pesticides can harm both the environment and the beneficial organisms in your garden. Instead, opt for natural pest management techniques, such as companion planting, attracting predatory insects, and using physical barriers like row covers. Additionally, maintaining a healthy garden through proper plant selection and care can help to prevent pest problems in the first place.

Renewable Energy Sources: Integrating renewable energy sources into your permaculture garden can help to reduce your reliance on fossil fuels and minimize your ecological footprint. Solar panels, wind turbines, and small-scale hydroelectric systems are all viable options for generating clean, sustainable energy. Additionally, consider using energy-efficient tools and appliances, such as solar-powered water pumps or LED lighting.

Waste Reduction and Recycling: In a permaculture garden, waste is viewed as a valuable resource rather than something to be discarded. Implementing strategies such as composting, worm farming, and using plant-based mulches can help recycle organic waste into the garden. Additionally, consider repurposing materials like pallets, tires, or old containers for use in garden structures or as planters.

By incorporating these sustainable techniques and systems into your permaculture garden design, you can create a resilient, productive, and environmentally-friendly space that will thrive for years to come. Remember, the key to successful permaculture is to observe and learn from nature, adapting and refining your practices as your garden evolves.

Creating a Maintenance Plan for Long-Term Success

A flourishing permaculture garden is a living testament to the power of sustainable design and ecological harmony. However, even the most well-designed garden requires ongoing care and attention to ensure its long-term success. This section will explore the essential components of a maintenance plan that will keep your permaculture garden thriving for years to come.

First and foremost, it's crucial to remember that a permaculture garden is a **dynamic ecosystem** constantly evolving and adapting to changing conditions. As such, your maintenance plan should be flexible and responsive to your garden's needs. Regular observation and interaction with your garden will help you identify any issues that may arise and address them promptly.

One of the critical aspects of permaculture maintenance is managing the growth and spread of plants. This includes pruning, thinning, and harvesting to ensure your plants remain healthy and productive. For example, fruit trees may require annual pruning to encourage new growth and prevent overcrowding, while perennial plants may need to be divided and replanted to maintain vigor. Regularly harvesting your garden's bounty not only provides you with fresh, nutritious food but also helps to keep plants in check and prevent them from becoming too dominant.

Another vital component of a permaculture maintenance plan is soil health. Healthy soil is the foundation of a thriving garden, and monitoring and maintaining its fertility and structure is essential. This can be achieved through regular mulching, composting, and the use of cover crops to protect and nourish the soil. Additionally, rotating your crops and incorporating a diverse range of plants will help to prevent nutrient depletion and the buildup of pests and diseases.

Pest and disease management is also an important aspect of permaculture maintenance. In a well-designed garden, natural predators and beneficial insects will help to keep pests in check. However, it's still essential to monitor your garden for signs of infestation or disease and take appropriate action when necessary. This may include introducing additional beneficial insects, using organic pest control methods, or adjusting your planting strategies to reduce the risk of disease.

Water management is another critical factor in maintaining a successful permaculture garden. Ensuring that your plants receive the right amount of water – not too much or too little – is crucial for their health and productivity. This may involve adjusting your irrigation systems, collecting rainwater, or implementing water-saving techniques such as mulching and planting drought-tolerant species.

Finally, a long-term maintenance plan should also consider the ongoing evolution and development of your permaculture garden. As your garden matures, you may need to reassess its design and make adjustments to accommodate changing conditions or new goals. This could involve introducing new plants or animals, expanding your garden's size, or incorporating additional sustainable techniques and systems.

In conclusion, creating a maintenance plan for your permaculture garden ensures its long-term success. By regularly monitoring and caring for your garden, you can enjoy the many benefits of a thriving, sustainable ecosystem that provides nourishment, beauty, and inspiration for years to come.

Bringing Your Permaculture Garden to Life

As we reach the end of this chapter, it is time to take a step back and appreciate the incredible journey you have embarked upon in designing your permaculture garden. By now, you should have a solid understanding of the principles and techniques that will help you create a thriving, sustainable, and resilient ecosystem right in your backyard.

Throughout this chapter, we have explored the importance of assessing your site and its resources, selecting the right plants and animals for your garden, integrating sustainable techniques and systems, and creating a maintenance plan for long-term success. Each of these elements plays a crucial role in bringing your permaculture garden to life.

As you begin to implement your design, remember that **permaculture is not a one-size-fits-all approach.** It is a dynamic and adaptive process that encourages experimentation and learning from experience. Embrace the challenges and surprises that may arise along the way, as they will only deepen your understanding of the intricate relationships within your garden's ecosystem.

One of the most rewarding aspects of permaculture is witnessing the transformation of your garden as it evolves into a self-sustaining, productive, and biodiverse haven. With time, patience, and dedication, you will see the fruits of your labor as your garden becomes a living testament to the power of permaculture principles.

As you nurture your permaculture garden, you cultivate a healthier and more sustainable environment for yourself and your community and contribute to the **global movement** toward a more resilient and regenerative future. Your garden is a small but significant piece of the larger puzzle, and your efforts will inspire others to embark on their own permaculture journey.

In conclusion, designing and bringing your permaculture garden to life is exciting and fulfilling. As you continue to learn, adapt, and grow alongside your garden, you will undoubtedly experience the profound satisfaction of living in harmony with nature. So, go forth and cultivate your permaculture paradise, knowing that you are making a lasting, positive impact on the world around you.

Chapter Summary

- Permaculture is a holistic approach to gardening that focuses on creating sustainable, self-sufficient systems that work in harmony with nature.
- Assessing your site and its resources is crucial for understanding your garden's unique features and resources, which will influence its design.
- Choosing the right plants and animals for your garden is essential for creating a thriving ecosystem that benefits both you and the environment.
- Integrating sustainable techniques and systems, such as water conservation, soil health, and natural pest

management, is key to creating a resilient and environmentally-friendly permaculture garden.

- A maintenance plan is necessary to ensure your permaculture garden's long-term success, including managing plant growth, soil health, pest and disease management, and water management.
- Permaculture is a dynamic and adaptive process that encourages experimentation and learning from experience, allowing for adjustments and improvements over time.
- Witnessing the transformation of your garden into a self-sustaining, productive, and biodiverse haven is one of the most rewarding aspects of permaculture.
- By cultivating a permaculture garden, you contribute to a global movement towards a more resilient and regenerative future, inspiring others to embark on their own permaculture journey.

3

SOIL HEALTH AND REGENERATION TECHNIQUES

A thriving garden with a variety of colorful plants and flowers, where the soil is visibly rich and healthy, with earthworms and microorganisms working together to regenerate the soil.

In the world of permaculture, the foundation of a thriving ecosystem lies in the health of its soil. Soil health and regeneration are essential components of sustainable agriculture and play a critical role in maintaining the balance of life on Earth. This chapter delves into the fascinating world of soil health and regeneration, exploring its importance in permaculture and providing practical techniques to nurture and restore this vital resource.

Soil is much more than just dirt; it is a complex, living ecosystem teeming with microorganisms, nutrients, and minerals that support plant growth and provide a habitat for countless organisms. Healthy soil is the backbone of a productive permaculture garden, as it supports the growth of diverse plant species, aids in water retention, and helps to regulate the climate. Unfortunately, conventional agricultural practices have led to widespread soil degradation, resulting in reduced fertility, erosion, and the loss of valuable topsoil.

In response to these challenges, permaculture offers a holistic approach to soil health and regeneration that emphasizes working with nature rather than against it. By understanding the intricacies of soil composition and implementing regenerative techniques, permaculture practitioners can create a thriving, resilient ecosystem that supports both human needs and the environment.

In this chapter, we will explore the importance of soil composition and its role in supporting a healthy permaculture system. We will discuss natural soil regeneration techniques that can be easily incorporated into your permaculture garden, as well as specific permaculture practices that promote soil health. Finally, we will provide guidance on monitoring and maintaining soil health, ensuring the lasting success of your permaculture endeavors.

So, let us embark on this journey to better understand the complex world beneath our feet and learn how to harness the power of healthy soil to create a thriving, sustainable permaculture garden.

Understanding Soil Composition and Its Importance

Soil is the foundation of any thriving permaculture garden. Therefore, understanding its composition is crucial for creating a sustainable and productive ecosystem. In this section, we will delve into the components of soil, the roles they play in supporting plant life, and why maintaining a healthy soil composition is essential for successful permaculture.

Soil is a complex and dynamic mixture of organic matter, minerals, water, air, and countless microorganisms. These components work together to create a living, breathing ecosystem that supports plant growth and contributes to the overall health of our environment. So let's take a closer look at each component and its role in **soil health.**

Organic matter: This includes decomposing plant and animal material, such as leaves, branches, and manure. As these materials break down, they release essential nutrients into the soil, which plants can then absorb through their roots. Organic matter also helps to improve soil structure, making it more porous and allowing air and water to circulate more easily.

Minerals: Soil contains various minerals, such as sand, silt, and clay, which provide essential nutrients for plant growth. The proportions of these minerals in the soil determine its texture, which in turn affects its ability to retain water and nutrients. A well-balanced soil composition will have a mix of all three mineral types, providing an ideal environment for plant growth.

Water: Soil moisture is vital for plant growth, as it dissolves and transports nutrients to the plant roots. Water also plays a crucial role in maintaining the soil's structure. It helps bind particles together and creates a stable environment for plant roots to anchor themselves.

Air: Oxygen is essential for the survival of both plants and soil-dwelling microorganisms. A well-aerated soil allows plant roots to access the oxygen they need for respiration while also providing a hospitable environment for beneficial microbes to thrive.

Microorganisms: Soil is teeming with billions of microscopic organisms, including bacteria, fungi, and protozoa. These tiny creatures play a vital role in breaking down organic matter, recycling nutrients, and maintaining soil structure. They also help to protect plants from diseases and pests by competing with harmful organisms and producing natural antibiotics.

Understanding the importance of each component in soil composition is the first step toward creating a healthy and productive permaculture garden. By nurturing and maintaining a well-balanced soil ecosystem, you can ensure your plants have access to the nutrients, water, and air they need to grow strong and resilient. This will lead to a more sustainable and self-sufficient garden, with minimal need for external inputs and reduced reliance on synthetic fertilizers and pesticides. In the next section, we will explore natural soil regeneration techniques that can help you achieve this goal.

Natural Soil Regeneration Techniques

In this section, we will delve into the world of natural soil regeneration techniques that can be easily incorporated into your permaculture garden. These methods not only help improve soil health but also contribute to the overall sustainability of your garden's ecosystem. By understanding and implementing these techniques, you will be taking a significant step towards creating a thriving and productive permaculture landscape.

Mulching: Mulching is the process of covering the soil surface with organic materials such as straw, leaves, wood chips, or even cardboard. This layer of mulch serves multiple purposes: it helps retain moisture, suppresses weed growth, regulates soil temperature, and gradually decomposes to add nutrients back into the soil. Additionally, mulching encourages the presence of beneficial organisms like earthworms, which aid in soil aeration and nutrient cycling.

Composting: Composting is the controlled decomposition of organic matter, resulting in a nutrient-rich, humus-like substance called compost. Adding compost to your garden provides your plants with essential nutrients and improves soil structure. Composting also helps recycle kitchen scraps and garden waste, reducing the amount of waste sent to landfills.

Cover cropping: Cover crops are plants grown specifically to improve soil health and fertility. They are typically fast-growing and can be easily incorporated into the soil after they have served their purpose. Some common cover crops include legumes (such as clover and beans), grasses (like rye and oats), and brassicas (like mustard and radish). Cover crops help prevent soil erosion, suppress weeds, and increase organic matter content. Legume cover crops also have the added benefit of fixing nitrogen from the atmosphere, making it available for other plants to use.

No-till or low-till gardening: Traditional tilling methods can disrupt soil structure, harm beneficial organisms, and contribute to soil erosion. No-till or low-till gardening involves minimal soil disturbance, allowing it to maintain its structure and supporting a healthy ecosystem of microorganisms. This can be achieved using techniques such as sheet mulching, lasagna gardening, or simply planting directly into the soil without turning it over.

Crop rotation: Crop rotation is the practice of growing different types of plants in the same area over a series of seasons. This helps prevent the buildup of pests and diseases, as well as the depletion of specific nutrients in the soil. By rotating crops, you can ensure that your soil remains balanced and fertile, giving your plants the nutrients they need to thrive.

Plant diversity: Incorporating a wide variety of plants in your permaculture garden not only creates a visually appealing landscape but also contributes to soil health. Different plants have varying root depths and nutrient requirements, which can help maintain a balanced soil ecosystem. Additionally, diverse plantings attract a range of beneficial insects and pollinators, further supporting the overall health of your garden.

In conclusion, natural soil regeneration techniques are essential for a successful permaculture garden. By implementing these practices, you will be nurturing a healthy, fertile, and sustainable soil ecosystem that will support the growth and productivity of your plants for years to come.

Implementing Permaculture Practices for Soil Health

In this section, we will delve into the practical implementation of permaculture practices that can significantly improve soil health. By incorporating these techniques into your garden or farm, you will create a sustainable and thriving ecosystem that supports plant and animal life. Let's explore some of the most effective permaculture practices for soil health.

No-Till Gardening: One of the core principles of permaculture is to work with nature, not against it. Traditional agricultural practices often involve tilling the soil, which disrupts the natural structure and balance of the soil ecosystem. No-till gardening is a permaculture practice that involves planting directly into the soil without turning it over. This helps maintain the soil's structure, preserves beneficial microorganisms, and reduces erosion.

Mulching: Mulching is the practice of covering the soil surface with organic materials such as leaves, straw, or wood chips. This serves multiple purposes: it helps retain moisture, suppresses weed growth, and provides a habitat for beneficial organisms. As the mulch breaks down, it also adds valuable nutrients back into the soil, improving its fertility and structure.

Composting: Composting is the process of decomposing organic matter into a nutrient-rich, soil-like material called humus. Adding compost to your garden provides your plants with essential nutrients and improves the overall soil structure. Composting also helps to reduce waste, as it repurposes kitchen scraps and yard waste that would otherwise end up in a landfill.

Plant Diversity: A diverse range of plant species is essential for a healthy permaculture garden. Different plants have varying root structures and nutrient requirements, which can help improve soil structure and nutrient cycling. Additionally, a diverse garden is more resilient to pests and diseases, reducing the need for chemical interventions.

Crop Rotation: Crop rotation is the practice of growing different types of crops in the same area over a series of seasons. This helps to prevent the buildup of pests and diseases that can occur when the same crop is grown repeatedly in the exact location. Crop rotation also helps to maintain soil fertility, as different plants have different nutrient requirements and can help balance the soil's nutrient levels.

Cover Crops: Cover crops are plants that are grown primarily to improve soil health rather than for harvest. They can be used to protect the soil from erosion, suppress weeds, and add organic matter to the soil. Some cover crops, such as legumes, can also fix nitrogen from the atmosphere, improving soil fertility.

Planting Perennials: Perennial plants are those that live for more than two years, and they play a vital role in permaculture gardens. Their deep root systems help to improve soil structure and reduce erosion. Additionally, perennials require less maintenance than annuals, as they do not need to be replanted each year.

By implementing these permaculture practices in your garden or farm, you will be well on your way to creating a healthy, regenerative soil ecosystem. In the next section, we will discuss monitoring and maintaining soil health in your permaculture garden to ensure its long-term success.

Monitoring and Maintaining Soil Health in Your Permaculture Garden

A thriving permaculture garden is a testament to the harmony between nature and human intervention. One of the critical aspects of this symbiosis is the health of the soil that supports the entire ecosystem. Monitoring and maintaining soil health is an **ongoing process.** It requires a keen understanding of the soil's needs and the ability to adapt to its ever-changing conditions. In this section, we will explore the various ways you can monitor and maintain the health of your permaculture garden's soil.

To begin with, it is essential to establish a baseline for your soil's health. This can be done by conducting a soil test, which will provide valuable information about the soil's pH, nutrient levels, and organic matter content. Soil tests can be done using at-home testing kits or by sending a sample to a professional laboratory. By understanding your soil's current state, you can make informed decisions about improving and maintaining its health.

Once you have established your soil's baseline, it is crucial to monitor its health regularly. This can be done through visual inspections, where you look for signs of healthy soil, such as the presence of earthworms, a crumbly texture, and a rich, earthy smell. Additionally, you can perform periodic soil tests to track changes in nutrient levels and pH over time.

One of the critical aspects of maintaining soil health in a permaculture garden is ensuring that the soil remains well-aerated and has proper drainage. Compacted soil can lead to poor root growth and reduced nutrient uptake by plants. To prevent soil compaction, avoid walking on your garden beds and use mulch to create pathways. You can also use tools like broad forks to gently aerate the soil without disrupting its structure.

Another essential aspect of soil health is maintaining a balance of nutrients. In a permaculture garden, this is often achieved through the use of organic matter, such as compost, manure, and mulch. These materials not only provide essential nutrients for plant growth but also help to improve the soil's structure and water retention capabilities. Regularly adding organic matter to your garden beds will **ensure your soil remains fertile** and productive.

In addition to organic matter, cover crops can play a vital role in maintaining soil health. Cover crops, such as clover, vetch, and rye, help to prevent soil erosion, suppress weeds, and fix nitrogen in the soil. By incorporating cover crops into your garden's rotation, you can ensure that your soil remains healthy and nutrient-rich.

Lastly, it is essential to practice crop rotation in your permaculture garden. By rotating the types of plants grown in a specific area, you can help prevent the buildup of pests and diseases while ensuring that the soil's nutrients are not depleted. A well-planned crop rotation will contribute to the overall health and productivity of your garden's soil.

In conclusion, monitoring and maintaining soil health is vital to successful permaculture gardening. By understanding your soil's needs and implementing practices such as proper aeration, organic matter addition, cover cropping, and crop rotation, you can ensure that your garden's soil remains healthy and productive for years to come. Remember, healthy soil is the foundation of a thriving permaculture garden. Your efforts in nurturing it will be rewarded with **bountiful harvests** and a flourishing ecosystem.

The Lasting Impact of Healthy Soil on Permaculture

In conclusion, the vitality and sustainability of any permaculture system are deeply rooted in the health of its soil. As we have explored throughout this chapter, understanding and nurturing soil composition is of paramount importance for the long-term success of your permaculture garden. By implementing the soil health and regeneration techniques discussed, you will not only witness a thriving ecosystem but also contribute to the global movement toward sustainable agriculture and environmental preservation.

The lasting impact of healthy soil on permaculture is multifaceted, as it not only supports the growth of diverse and resilient plant life but also fosters a balanced ecosystem teeming with beneficial microorganisms and insects. This, in turn, leads to increased productivity, reduced need for external inputs, and a more self-sufficient garden that can withstand the test of time.

Moreover, the principles of permaculture extend beyond the confines of your garden, as they encourage a shift in mindset towards a more holistic and interconnected view of the world around us. By nurturing the soil beneath our feet, we also nurture our communities, our planet, and ourselves. The regeneration of soil health is not just a means to an end but a journey of discovery and connection with the natural world.

As you continue on your permaculture journey, remember that soil health is an ongoing process that requires consistent care and attention. Regularly monitoring and maintaining your soil's health will ensure that it remains a fertile and productive foundation for your permaculture garden for years to come. Embrace the challenges and rewards of this responsibility, and take pride in the knowledge that your efforts contribute to a greener, more sustainable future for all.

In the words of the renowned environmentalist and permaculture pioneer, *Bill Mollison, "Permaculture is a philosophy of working with, rather than against nature; of protracted and thoughtful observation rather than protracted and thoughtless labor; and of looking at plants and animals in all their functions, rather than treating any area as a single project system."*

By prioritizing soil health and regeneration, you embody this philosophy and play a vital role in the global movement towards a more sustainable and harmonious existence with the natural world.

Chapter Summary

- Soil health is the foundation of a thriving permaculture ecosystem, playing a critical role in maintaining the balance of life on Earth.
- Soil is a complex, living ecosystem consisting of organic matter, minerals, water, air, and microorganisms, all of which contribute to plant growth and overall environmental health.
- Conventional agricultural practices have led to widespread soil degradation, making it essential to adopt permaculture techniques that prioritize soil health and regeneration.
- Natural soil regeneration techniques, such as mulching, composting, cover cropping, no-till gardening, crop rotation, and plant diversity, can significantly improve soil health and contribute to a sustainable garden ecosystem.
- Permaculture practices, including no-till gardening, mulching, composting, plant diversity, crop rotation, cover crops, and planting perennials, can help maintain and improve soil health.

- Monitoring and maintaining soil health is an ongoing process that involves regular visual inspections, soil testing, proper aeration, organic matter addition, cover cropping, and crop rotation.
- Healthy soil has a lasting impact on permaculture, supporting diverse and resilient plant life, fostering a balanced ecosystem, and contributing to the global movement toward sustainable agriculture and environmental preservation.
- Embracing the principles of permaculture and prioritizing soil health is essential for creating a greener, more sustainable future for all.

- Monitoring and maintaining soil health is an ongoing process that involves regular visual inspections, soil testing, proper aeration, organic matter addition, cover cropping, and crop rotation.
- Healthy soil has a lasting impact on permaculture, supporting diverse and resilient plant life, fostering a balanced ecosystem, and contributing to the global movement toward sustainable agriculture and environmental preservation.
- Embracing the principles of permaculture and prioritizing soil health is essential for creating a greener, more sustainable future for all.

4

WATER MANAGEMENT AND CONSERVATION

A futuristic city with advanced water management systems, including rainwater harvesting skyscrapers, underground water storage facilities, and smart irrigation networks in urban gardens.

In the realm of permaculture, water is a vital and indispensable resource that plays a crucial role in the overall health and productivity of an ecosystem. As a finite resource, it is essential to manage and conserve water effectively to ensure the sustainability and resilience of our gardens, farms, and communities. This chapter will delve into the various aspects of water management and conservation in permaculture, providing practical solutions and techniques that can be applied to any landscape or environment.

As a holistic design system, permaculture emphasizes the importance of working with nature and its patterns rather than against it. This principle is particularly relevant when it comes to water management and conservation, as the natural water cycle is a perfect example of a closed-loop system that efficiently distributes and recycles water. By understanding and mimicking these natural processes, we can create regenerative and resilient systems that not only conserve water but also enhance the overall health of the ecosystem.

Water management and conservation in permaculture involve a multifaceted approach that includes harvesting rainwater, designing efficient irrigation systems, implementing greywater and rainwater systems, and employing strategies to reduce water waste and evaporation. By integrating these techniques into our permaculture designs, we can create landscapes that are not only productive and beautiful but also water-wise and sustainable.

In the following sections, we will explore the importance of water harvesting techniques and how they can be implemented in various settings. We will also discuss the design and implementation of efficient irrigation systems that minimize water waste and maximize productivity. Additionally, we will delve into the world of greywater and rainwater systems, providing practical solutions for capturing and reusing water in our homes and gardens.

Finally, we will examine strategies for reducing water waste and evaporation, ensuring that every drop of water is used effectively and responsibly.

As we navigate through this chapter, it is important to remember that **water is a precious resource** that connects all living beings on our planet. By adopting responsible water management and conservation practices in our permaculture designs, we are not only contributing to the health and vitality of our ecosystems but also playing a vital role in the global effort to preserve and protect our planet's most valuable resource.

The Importance of Water Harvesting Techniques

In the realm of permaculture, water is undeniably one of the most vital resources for sustaining life and promoting the growth of plants, animals, and microorganisms. As a result, it is crucial for practitioners to develop and implement effective water harvesting techniques that not only ensure an adequate supply of water but also contribute to the overall health and sustainability of the ecosystem. This section will delve into the importance of water harvesting techniques and their role in fostering a thriving permaculture landscape.

Water harvesting techniques are essential for several reasons. Firstly, they help capture and store water from various sources such as rainfall, runoff, and underground springs. By doing so, these techniques provide a reliable and consistent water supply for plants and animals, even during drought or water scarcity. This is particularly important in arid and semi-arid regions, where water is often a limiting factor for the success of permaculture systems.

Secondly, water harvesting techniques are crucial in preventing soil erosion and nutrient loss. When water is allowed to flow freely across the landscape, it can carry away valuable topsoil and essential nutrients, leaving behind a degraded and less fertile environment. By capturing and directing water through carefully designed systems, permaculture practitioners can **minimize erosion** and ensure nutrients are retained within the ecosystem.

Moreover, water harvesting techniques contribute to the overall health and resilience of the permaculture system by promoting biodiversity and supporting the growth of various plant and animal species. For instance, creating swales, ponds, and other water features can provide a habitat for aquatic plants and animals and create microclimates that support a diverse range of species. This, in turn, leads to a more robust and adaptable ecosystem that can better withstand external pressures such as climate change and invasive species.

Additionally, water harvesting techniques can reduce the reliance on external water sources, such as municipal water supplies or groundwater extraction. By capturing and utilizing water on-site, permaculture practitioners can minimize their environmental footprint and contribute to conserving valuable water resources. This is particularly important in areas where water scarcity is a growing concern, as it helps to promote a more sustainable and self-sufficient approach to water management.

In conclusion, water harvesting techniques are indispensable to any successful permaculture system. By capturing, storing, and distributing water efficiently and sustainably, these techniques ensure the availability of this precious resource and contribute to the ecosystem's overall health, resilience, and sustainability. As we continue to face the challenges of climate change and increasing water scarcity, the importance of effective water harvesting techniques in permaculture cannot be overstated.

Designing Efficient Irrigation Systems

Designing efficient irrigation systems is crucial for water management and conservation in permaculture. With the right approach, you can ensure that your plants receive the optimal amount of water while minimizing waste and preserving this precious resource. In this section, we will explore the fundamental principles of designing efficient irrigation systems and discuss some of the most popular techniques used by permaculture practitioners around the world.

The first step in designing an efficient irrigation system is to understand the specific needs of your plants and the characteristics of your site. Different plants have varying water requirements, and factors such as soil type, climate, and topography can significantly impact the effectiveness of your irrigation system. By conducting a thorough site analysis and researching the water needs of your plants, you can tailor your irrigation system to suit the unique conditions of your permaculture garden.

One of the core principles of efficient irrigation is to deliver water directly to the root zone of your plants. This minimizes evaporation and runoff, ensuring your plants absorb the water you provide. **Drip irrigation**, also known as trickle irrigation, is a popular method that achieves this goal. In a drip irrigation system, water is slowly released through a network of tubes and emitters, allowing it to seep directly into the soil around the roots of your plants. This method is highly efficient, as it minimizes water waste and reduces the risk of overwatering.

Another effective irrigation technique is the use of **swales**, which are shallow-level trenches dug along the contour lines of your landscape. Swales help capture and slow the water flow, allowing it to infiltrate the soil and provide moisture to your plants. By strategically placing swales throughout your permaculture site, you can create a passive irrigation system that makes the most of the natural water movement across your land.

In addition to choosing a suitable irrigation method, it's essential to consider the timing and frequency of your watering. Watering your plants early in the morning or late in the evening can help to reduce evaporation, as the temperatures are cooler and the sun is less intense. Additionally, it's important to monitor the moisture levels in your soil and adjust your watering schedule accordingly. Overwatering can be as detrimental to your plants as underwatering, so finding the right balance for your site is crucial.

Finally, **mulching** is an excellent strategy for improving the efficiency of your irrigation system. By covering the soil around your plants with a layer of organic material, such as straw, leaves, or wood chips, you can help to retain moisture, regulate soil temperature, and reduce evaporation. Mulching also has the added benefit of suppressing weeds and adding nutrients to the soil as it breaks down over time.

In conclusion, designing efficient irrigation systems is essential to water management and conservation in permaculture. By understanding the needs of your plants and the characteristics of your site, selecting appropriate irrigation methods, and implementing strategies to reduce water waste and evaporation, you can create a sustainable and productive permaculture garden that thrives with minimal input. As water scarcity becomes an increasingly pressing global issue, the importance of efficient irrigation systems in permaculture will only continue to grow.

Implementing Greywater and Rainwater Systems

In the quest for sustainable living and efficient water management, greywater and rainwater systems have emerged as vital components of permaculture practices. These systems help conserve water and contribute to the overall health of your garden and the environment. In this section, we will delve into the benefits of these systems and provide practical guidance on how to implement them in your permaculture design.

Greywater refers to the gently used water from sinks, showers, and washing machines that can be reused for irrigation purposes. Recycling this water can significantly reduce your reliance on freshwater sources and lower your water bills. Rainwater, however, is a **free and abundant resource** that can be harvested and stored for various uses, including irrigation and household use, when properly treated.

To implement a greywater system, you must first assess the quality and quantity of the water you generate. This will help you determine the most suitable plants for your garden and the necessary filtration methods. It is essential to avoid using water contaminated with harsh chemicals, as it can harm your plants and the soil. Opt for eco-friendly cleaning products and consider installing a simple filtration system, such as a mulch basin, to remove any remaining particles before the water reaches your plants.

Next, you will need to design a distribution system that evenly disperses the greywater throughout your garden. Drip irrigation and subsurface systems are ideal, as they deliver water directly to the plant roots and minimize evaporation. Be sure to check your local regulations, as some areas may have specific guidelines or restrictions on greywater use.

Rainwater harvesting, on the other hand, involves collecting and storing rainwater for later use. This can be achieved through various methods, such as installing gutters and downspouts on your roof, which direct the water into storage tanks or barrels. You can also create swales and rain gardens to capture and store rainwater directly in your landscape.

When designing your rainwater system, consider factors such as the size of your roof, the average rainfall in your area, and your water needs. This will help you determine the appropriate storage capacity and whether you need additional filtration or treatment systems. Keep in mind that stored rainwater can be used for more than just irrigation; with proper treatment, it can also be used for household purposes, such as flushing toilets or even drinking.

In conclusion, implementing greywater and rainwater systems in your permaculture design is an effective way to conserve water, reduce waste, and promote a healthier ecosystem. By thoughtfully planning and incorporating these systems into your landscape, you can create a more sustainable and self-sufficient environment that benefits both your garden and the planet.

Strategies for Reducing Water Waste and Evaporation

In the world of permaculture, water is a precious resource that must be used wisely and efficiently. As we've explored various water management techniques in this chapter, it's crucial to address strategies for reducing water waste and evaporation. By implementing these methods, you can ensure that your permaculture garden remains sustainable and productive while also minimizing your ecological footprint.

Proper **irrigation scheduling** is one of the most effective ways to reduce water waste. By watering your plants at the right time and in the right amounts, you can avoid overwatering and ensure that your plants receive the moisture they need without wasting water. To achieve this, monitor the moisture levels in your soil and adjust your irrigation schedule accordingly. Watering in the early morning or late evening can also help minimize evaporation, as the temperatures are more relaxed and the sun is less intense.

Mulching is another essential strategy for reducing water waste and evaporation. Covering the soil with a layer of organic material, such as straw, wood chips, or leaves, can help retain moisture, regulate soil temperature, and reduce evaporation. Mulch also helps suppress weeds, which compete with your plants for water and nutrients. Be sure to replenish your mulch regularly, as it will break down over time and contribute to the fertility of your soil.

Incorporating **shade** into your permaculture design can also help reduce water waste and evaporation. By planting taller plants or trees to shade smaller plants, you can create a microclimate that helps retain moisture and reduces the need for frequent watering. Additionally, consider using shade cloth or other materials to protect your plants from the intense sun during the hottest parts of the day.

Another strategy for reducing water waste is to select **drought-tolerant plants** for your permaculture garden. These plants have adapted to thrive in low-water conditions. As a result, they can help you conserve water while enjoying a productive and diverse garden. Some examples of drought-tolerant plants include lavender, rosemary, sage, and various succulents. Be sure to research the specific water requirements of each plant to ensure that you provide the appropriate amount of moisture for their needs.

Lastly, proper **garden maintenance** can also help reduce water waste and evaporation. Regularly pruning your plants promotes healthy growth and reduces the amount of water they require. Additionally, keeping your garden free of weeds and pests can help ensure that your plants receive the water they need without competition from unwanted invaders.

In conclusion, implementing these strategies for reducing water waste and evaporation can help you create a more sustainable and efficient permaculture garden. By carefully managing your water resources and making thoughtful choices in your garden design, you can contribute to a healthier planet and enjoy the many benefits of a thriving permaculture ecosystem.

The Future of Water Conservation in Permaculture

As we reach the end of our exploration into water management and conservation in permaculture, it is essential to consider the future of this vital aspect of sustainable living. With climate change and increasing water scarcity becoming pressing global issues, efficient and environmentally friendly water management practices are more **crucial** than ever. Permaculture, with its emphasis on working in harmony with nature, offers a promising solution to these challenges.

In the coming years, we expect to see a greater focus on water conservation within the permaculture community. This will likely involve developing and refining innovative water harvesting techniques and the widespread adoption of efficient irrigation systems. In addition, as more people recognize the benefits of permaculture, we can anticipate a surge in the implementation of greywater and rainwater systems, which will not only conserve water but also reduce the strain on municipal water supplies.

Moreover, the future of water conservation in permaculture will likely involve a greater emphasis on education and awareness-raising. As more individuals understand the importance of water conservation and its role in creating a sustainable future, they will be more likely to adopt permaculture practices in their lives. This, in turn, will lead to a ripple effect as the benefits of water conservation become increasingly apparent on a larger scale.

Another critical aspect of the future of water conservation in permaculture is the ongoing development of strategies for reducing water waste and evaporation. This will involve not only the creation of new techniques but also the refinement of existing ones, ensuring that they are as effective as possible. By continually improving upon these strategies, permaculture practitioners will be better equipped to conserve water and create resilient ecosystems.

In conclusion, the future of water conservation in permaculture is bright, filled with innovation, collaboration, and a commitment to working in harmony with the natural world. By embracing permaculture principles and implementing effective water management practices, we can help create a more sustainable and resilient future for ourselves and future generations. As we continue to face the challenges of climate change and water scarcity, the importance of water conservation in permaculture will only continue to grow, making it an essential component of our efforts to build a better world.

Chapter Summary

- Water is a vital and indispensable resource in permaculture, playing a crucial role in the overall health and productivity of an ecosystem. Effective water management and conservation are essential for ensuring the sustainability and resilience of our gardens, farms, and communities.
- Water harvesting techniques are crucial for capturing and storing water from various sources, preventing soil erosion and nutrient loss, promoting biodiversity, and reducing reliance on external water sources.
- Designing efficient irrigation systems involves understanding the specific needs of plants and site characteristics, delivering water directly to the root zone, using methods such as drip irrigation and swales, and proper scheduling to minimize water waste and evaporation.
- Implementing greywater and rainwater systems can help conserve water, reduce waste, and promote a healthier ecosystem. Greywater systems recycle gently used water from sinks, showers, and washing machines for irrigation purposes, while rainwater systems collect and store rainwater for various uses.
- Reducing water waste and evaporation can be achieved through proper irrigation scheduling, mulching, incorporating shade into the garden design, selecting

drought-tolerant plants, and maintaining the garden properly.
- The future of water conservation in permaculture will likely involve a greater focus on innovative water harvesting techniques, efficient irrigation systems, and widespread adoption of greywater and rainwater systems.
- Education and awareness-raising will play a significant role in the future of water conservation in permaculture as more individuals understand the importance of water conservation and adopt permaculture practices in their own lives.
- The ongoing development of strategies for reducing water waste and evaporation, as well as the refinement of existing techniques, will be crucial in ensuring effective water conservation and creating resilient ecosystems in permaculture.

5

PLANT SELECTION AND COMPANION PLANTING

An enchanting garden scene where a tall sunflower stands proudly among a group of low-growing, vibrant green basil plants, with ladybugs happily exploring the leaves, showcasing the beauty and benefits of companion planting.

Welcome to chapter 5, where we delve into the fascinating world of plant selection and companion planting in the context of practical permaculture. As a permaculture enthusiast, you are likely already aware of the importance of working with nature to create sustainable, resilient, and productive systems. One of the critical aspects of achieving this is through the **careful selection of plants** and the **strategic placement** of these plants in your garden. This chapter aims to provide you with the knowledge and tools necessary to make informed decisions about the plants you choose and how they interact, ultimately contributing to the success of your permaculture garden.

Plant selection and companion planting are interconnected concepts crucial to designing and implementing any permaculture system. Plant selection refers to the process of choosing the right plants for your garden based on various factors such as climate, soil type, water availability, and the specific needs of your garden ecosystem. On the other hand, companion planting is the art and science of arranging plants in close proximity to one another so that they can mutually benefit from each other's presence. This can include providing nutrients, attracting beneficial insects, repelling pests, or creating a more favorable microclimate.

In this chapter, we will explore the importance of plant diversity in permaculture and the various criteria you should consider when selecting plants for your garden. We will then delve into the world of companion planting, discussing its benefits and techniques and providing examples of successful plant combinations. Finally, we will conclude with a discussion on how to maximize the potential of your permaculture garden through thoughtful plant selection and companion planting.

By the end of this chapter, you will have a deeper understanding of the principles and practices of plant selection and companion planting, empowering you to make informed decisions that will enhance the productivity, resilience, and sustainability of your permaculture garden. So, let's embark on this exciting journey together and discover the incredible potential that lies within the world of plants!

Understanding the Importance of Plant Diversity in Permaculture

Plant diversity is a fundamental principle of permaculture, and for a good reason. In this section, we will delve into the importance of plant diversity in permaculture, exploring its numerous benefits and how it contributes to a thriving, resilient, and sustainable garden ecosystem.

At the heart of permaculture is the concept of working with nature rather than against it. By embracing plant diversity, we are **mimicking the natural ecosystems** that have evolved over millions of years. In nature, a diverse range of plant species coexist, each fulfilling a specific role and contributing to the overall health and stability of the ecosystem. By incorporating a wide variety of plants in our permaculture gardens, we can harness the power of nature's wisdom and create a more resilient, productive, and sustainable system.

Key Benefits to Promoting Plant Diversity in Permaculture

Improved Soil Health: Different plants have varying root structures and nutrient requirements, which can help to improve soil structure and fertility. For example, deep-rooted plants can break up compacted soil and bring nutrients up to the surface, while nitrogen-fixing plants can replenish essential nutrients in the soil.

Pest and Disease Control: A diverse garden is less susceptible to pest and disease outbreaks, as it is more difficult for a single pest or pathogen to spread rapidly through a varied plant population. Additionally, many plants have natural pest-repelling properties and can be strategically placed to protect more vulnerable species.

Pollinator Attraction: A diverse range of flowering plants will attract a wide variety of pollinators, essential for reproducing many fruit and vegetable crops. This not only increases the productivity of your garden but also supports local biodiversity and the overall health of the ecosystem.

Climate Resilience: By incorporating plants with varying tolerances to temperature, moisture, and other environmental factors, you can create a garden that is more resilient to the impacts of climate change and extreme weather events.

Enhanced Productivity: Plant diversity can increase productivity, as different plants can be harvested at different times, ensuring a continuous supply of fresh produce throughout the growing season. Moreover, diverse gardens are more likely to include plants with multiple uses, such as those that provide food, medicine, or materials for building or crafting.

Aesthetics and Well-being: A diverse garden is not only functional but also beautiful and inspiring. Research has shown that spending time in natural environments with a high level of biodiversity can have significant mental health benefits, including reduced stress and increased well-being.

In conclusion, plant diversity is a cornerstone of permaculture, offering numerous benefits for both the garden and the gardener. By understanding and embracing the importance of plant diversity, we can create thriving, resilient, and sustainable permaculture systems that work in harmony with nature and provide abundant resources for ourselves and future generations.

Criteria for Choosing the Right Plants for Your Permaculture Garden

Selecting the right plants for your permaculture garden is crucial in ensuring its success. By carefully considering each plant's unique characteristics, you can create a productive and sustainable thriving ecosystem. This section will explore the critical criteria for choosing plants for your permaculture garden.

Climate and Hardiness

The first factor to consider when selecting plants is the climate in your garden. Choosing plants that are well-suited to the local climate is essential, as this will ensure their survival and growth. To do this, research the hardiness zones of various plants and select appropriate ones for your area. Hardiness zones are a classification system that indicates the minimum temperatures a plant can withstand.

Soil Type and Nutrient Requirements

Different plants have varying preferences for soil type and nutrient levels. Therefore, when selecting plants for your permaculture garden, it's important to consider the existing soil conditions and choose plants that will thrive in those conditions. Additionally, some plants can help improve soil quality by adding nutrients or breaking up compacted soil, so consider incorporating these into your garden design.

Sunlight and Water Needs

Just as plants have different soil preferences, they also have varying sunlight and water requirements. When choosing plants for your permaculture garden, consider the amount of sunlight and water that each plant will receive in its designated location. Select plants that are well-suited to the available resources, and arrange them in a way that maximizes their access to sunlight and water.

Size and Growth Habit

A plant's size and growth habits are essential factors to consider when designing your permaculture garden. Be sure to choose plants that will not outgrow their allotted space, as overcrowding can lead to competition for resources and reduced productivity. Additionally, consider the growth habits of each plant, such as whether they are climbers, groundcovers, or tall and upright, and arrange them in a way that maximizes their potential.

Edible and Medicinal Qualities

One of the primary goals of permaculture is to create a productive and self-sustaining ecosystem. To achieve this, it's essential to select plants that provide multiple functions, such as food and medicine. Therefore, when choosing plants for your permaculture garden, prioritize those that offer edible or medicinal benefits, as these will contribute to the overall productivity and resilience of your garden.

Pest and Disease Resistance

A healthy permaculture garden can withstand pests and diseases with minimal intervention. When selecting plants, prioritize those with natural pest and disease resistance. This will reduce the need for chemical treatments and help maintain a balanced ecosystem. Additionally, consider incorporating plants that attract beneficial insects, such as ladybugs and lacewings, which can help control pest populations.

Compatibility with Companion Plants

Finally, when choosing plants for your permaculture garden, consider their compatibility with other plants in the ecosystem. Companion planting is a technique that involves strategically placing plants together to maximize their mutual benefits, such as improved nutrient uptake, pest control, and pollination. You can create a harmonious and productive permaculture garden by selecting plants that work well together.

In conclusion, selecting the right plants for your permaculture garden involves careful consideration of various factors, including climate, soil type, sunlight and water needs, size and growth habits, edible and medicinal qualities, pest and disease resistance, and compatibility with companion plants. By keeping these criteria in mind, you can create a thriving and sustainable permaculture garden that meets your unique needs and goals.

The Art and Science of Companion Planting: Benefits and Techniques

Companion planting is a time-honored gardening technique practiced for centuries by farmers and gardeners alike. It is a method of growing different plants together so that they can mutually benefit from each other. This symbiotic relationship enhances the productivity and resilience of your permaculture garden and contributes to its overall health and sustainability. In this section, we will delve into the art and science of companion planting, exploring its benefits and techniques that can be applied to your permaculture garden.

The Benefits of Companion Planting

Companion planting offers a myriad of benefits for your permaculture garden. Some of the most notable advantages include the following:

Pest control: Certain plants can repel or attract insects, which can help keep pests in check without the need for harmful chemicals. For example, marigolds are known to deter nematodes, while basil can help ward off flies and mosquitoes.

Improved pollination: Planting flowers and herbs near your fruit and vegetable crops can attract pollinators such as bees and butterflies, which can improve the pollination and yield of your plants.

Enhanced nutrient availability: Some plants, like legumes, can fix nitrogen from the atmosphere and make it available to other plants. This can help improve soil fertility and reduce the need for synthetic fertilizers.

Natural support: Tall plants like sunflowers or corn can support climbing plants like beans and peas, reducing the need for artificial structures.

Improved soil structure: Planting deep-rooted plants alongside shallow-rooted ones can help break up compacted soil and improve its structure, allowing for better water infiltration and root growth.

The Techniques of Companion Planting

To successfully implement companion planting in your permaculture garden, consider the following techniques:

Research plant relationships: Before planting, take the time to research which plants grow well together and which ones may inhibit each other's growth. This will help you create a harmonious garden where plants can thrive and support one another.

Plan your garden layout: Design your garden in a way that takes advantage of the natural relationships between plants. For example, you can plant nitrogen-fixing legumes near nitrogen-hungry plants like tomatoes or corn or plant pest-repelling herbs near plants that are susceptible to pests.

Utilize plant guilds: Plant guilds are groups of plants that work well together and support each other's growth. For example, the *"Three Sisters"* guild consists of corn, beans, and squash, which provide support, nitrogen, and ground cover, respectively. Incorporating plant guilds into your permaculture garden can help create a balanced and productive ecosystem.

Rotate your crops: Rotating your crops each year can help prevent the buildup of pests and diseases in your garden. By changing the location of your plants, you can disrupt the life cycles of pests and reduce the likelihood of disease transmission.

Observe and adapt: As with any gardening technique, it's essential to observe your garden and make adjustments as needed. Pay attention to how your plants are growing and interacting with one another, and be prepared to change your companion planting strategy if necessary.

In conclusion, the art and science of companion planting can significantly enhance the productivity, health, and sustainability of your permaculture garden. By carefully selecting plants that complement one another and implementing thoughtful planting techniques, you can create a thriving ecosystem that maximizes the potential of your garden while minimizing the need for external inputs.

Maximizing the Potential of Your Permaculture Garden

In conclusion, the power to maximize the potential of your permaculture garden lies in the thoughtful and strategic implementation of plant selection and companion planting. When combined effectively, these two essential components can create a **thriving**, **sustainable**, and **productive ecosystem** that benefits the environment and enriches the lives of those who tend to it.

Throughout this chapter, we have explored the importance of plant diversity in permaculture and the various criteria to consider when choosing the right plants for your garden. By selecting plants well-suited to your climate, soil, and available resources, you can ensure that your garden remains resilient and adaptable to changing conditions.

We have also delved into the art and science of companion planting, highlighting the numerous benefits it offers, such as pest control, nutrient cycling, and improved pollination. By understanding the relationships between different plants and utilizing various companion planting techniques, you can create a harmonious and balanced garden that supports the growth and health of all its inhabitants.

As you embark on your permaculture adventure, remember that the key to success lies in observation, experimentation, and adaptation. By continually learning from your garden and adjusting your approach as needed, you can create a thriving ecosystem that not only provides for your needs but also contributes to the health and well-being of the planet.

In the end, the true beauty of permaculture lies in its ability to connect us with the natural world and remind us of our role as caretakers of the Earth. By embracing plant selection and companion planting, you are taking an important step towards creating a more sustainable, resilient, and abundant future for yourself and future generations.

Chapter Summary

- Plant selection and companion planting are essential components of permaculture, contributing to the success of a sustainable and productive garden ecosystem.
- Plant diversity is crucial in permaculture, as it promotes soil health, pest and disease control, pollinator attraction, climate resilience, enhanced productivity, and overall well-being.
- When selecting plants for a permaculture garden, consider factors such as climate, soil type, sunlight, water needs, size and growth habits, edible and medicinal qualities, pest and disease resistance, and compatibility with companion plants.
- Companion planting is the strategic arrangement of plants in close proximity to one another, allowing them to mutually benefit from each other's presence through improved nutrient uptake, pest control, and pollination.
- Benefits of companion planting include pest control, improved pollination, enhanced nutrient availability, natural support, and improved soil structure.
- Techniques for successful companion planting include researching plant relationships, planning the garden layout, utilizing plant guilds, rotating crops, and observing and adapting as needed.
- Maximizing the potential of a permaculture garden through plant selection and companion planting requires continuous observation, experimentation, and adaptation, ultimately contributing to a more sustainable, resilient, and abundant future.

6

CREATING DIVERSE AND RESILIENT ECOSYSTEMS

A lush tropical rainforest, highlighting the layers of the forest, from the canopy to the forest floor, and featuring a diverse array of flora and fauna interconnected within the ecosystem.

In the world of permaculture, the concepts of diversity and resilience are not just *buzzwords*; they are the very foundation of a thriving, sustainable ecosystem. As we face the challenges of climate change, habitat loss, and dwindling resources, it becomes increasingly important to create landscapes that can adapt and flourish in the face of adversity. In this chapter, we will explore the critical role that diverse and resilient ecosystems play in permaculture and how you can incorporate these principles into your own garden or farm.

At its core, permaculture is about **working with nature** rather than against it. This means observing and learning from the natural world and using that knowledge to design productive and self-sustaining systems. One of the key lessons we can learn from nature is the value of diversity. In a diverse ecosystem, a wide variety of plant and animal species coexist, each fulfilling a unique role and contributing to the overall health and stability of the system.

Diversity is important for several reasons. First, it provides a buffer against pests and diseases. When a single species dominates an ecosystem, it becomes vulnerable to outbreaks that can wipe out entire populations. By contrast, a diverse ecosystem is more resilient, as the presence of multiple species helps to keep pests and diseases in check.

Second, diversity promotes soil health and fertility. Different plants have different nutrient requirements and root structures, which can help to improve soil structure and prevent nutrient depletion. Finally, diversity supports a wide range of beneficial insects, birds, and other wildlife, which play crucial roles in pollination, pest control, and nutrient cycling.

Resilience is another crucial aspect of permaculture; it goes hand in hand with diversity. A resilient ecosystem can withstand disturbances like drought, floods, or temperature fluctuations and bounce back quickly. In a resilient system, the various components – plants, animals, and microorganisms – work together to maintain balance and stability, even in the face of change. This is particularly important in climate change, as we expect more frequent and severe weather events in the coming years.

In the following sections, we will delve deeper into the role of plant and animal species in building ecosystem diversity, discuss strategies for designing resilient permaculture systems, and explore how to integrate natural elements and processes into your garden or farm. So, let's embark on this exciting journey towards creating a more diverse and resilient world, one permaculture garden at a time!

The Role of Plant and Animal Species in Building Ecosystem Diversity

In the world of permaculture, diversity is critical to creating a thriving and resilient ecosystem. By incorporating a wide variety of plant and animal species, we can mimic the natural processes found in nature and promote a self-sustaining environment. In this section, we will delve into the importance of plant and animal species in building ecosystem diversity and how they contribute to the overall health and productivity of a permaculture system.

The Benefits of Plant Diversity

Plant diversity plays a crucial role in the overall health and productivity of a permaculture system. By incorporating a wide range of plant species, we can achieve several benefits:

Improved Soil Health: Different plants contribute to soil health in various ways. Some plants, such as legumes, fix nitrogen in the soil, while others, like deep-rooted plants, help break up compacted soil and improve aeration. This diverse array of plants helps create rich and fertile soil that supports the growth of other species.

Pest and Disease Control: A diverse plant community can help reduce the risk of pest and disease outbreaks. A single species dominating an area can create a monoculture more susceptible to pests and diseases. By incorporating a variety of plants, we can disrupt the life cycles of pests and diseases, making it more difficult for them to establish and spread.

Pollinator Attraction: A diverse array of flowering plants can attract a wide range of pollinators, such as bees, butterflies, and birds. These pollinators are essential for the reproduction of many plant species. They can increase the overall productivity of your permaculture system.

The Importance of Animal Species in Ecosystem Diversity

In addition to plant diversity, incorporating various animal species into your permaculture system can contribute to **ecosystem diversity.** Animals play several essential roles in a permaculture system, including:

Nutrient Cycling: Animals help cycle nutrients through the ecosystem by consuming plants and other organisms, breaking them down, and depositing their waste into the soil. This process helps maintain soil fertility and supports plant growth.

Pest Control: Many animal species, such as birds and insects, can help control pest populations by feeding on them. By encouraging a diverse range of animal species in your permaculture system, you can create a natural form of pest control that reduces the need for chemical interventions.

Pollination: As mentioned earlier, pollinators play a crucial role in plant reproduction. By incorporating animal species that act as pollinators, you can support the growth and productivity of your permaculture system.

Habitat Creation: By providing diverse habitats for various animal species, you can support a thriving ecosystem that contributes to the overall health and resilience of your permaculture system.

In conclusion, the role of plant and animal species in building ecosystem diversity is essential for the success of any permaculture system. By incorporating a wide range of species, we can create a self-sustaining environment resilient to pests, diseases, and environmental changes. The following sections will explore strategies for designing resilient permaculture systems and integrating natural elements and processes to further enhance ecosystem diversity.

Designing for Resilience: Strategies for Adapting to Environmental Changes

In a world where climate change and environmental degradation are increasingly affecting our ecosystems, designing resilient permaculture systems is more important than ever. By incorporating strategies that enable adaptation to environmental changes, we can create landscapes that withstand disturbances and thrive in adversity. This section will explore various strategies for designing resilient permaculture systems that adapt to various environmental changes.

Embrace diversity: One of the most effective ways to build resilience is by incorporating a diverse range of plant and animal species into your permaculture system. This diversity helps to create a more stable ecosystem, as different species can fill various ecological niches and provide a wide range of functions. In addition, in the event of a disturbance, such as a disease outbreak or pest infestation, a diverse ecosystem is more likely to recover quickly and maintain its overall health.

Utilize polycultures: Polycultures, or the practice of growing multiple species together, can help to increase the resilience of your permaculture system. By interplanting different species, you can create a more complex and interconnected ecosystem that is better able to withstand environmental changes. Polycultures can also help reduce the risk of disease and pest infestations, making it more difficult for pests to spread and establish themselves within the system.

Design for redundancy: Redundancy is critical in designing resilient systems. By incorporating multiple species that serve similar functions, you can ensure that your permaculture system continues functioning even if one species is lost or damaged. For example, planting several types of nitrogen-fixing plants can help to maintain soil fertility even if one species is affected by a disease or pest.

Foster soil health: Healthy soil is the foundation of a resilient permaculture system. By focusing on building and maintaining soil health, you can create an environment that supports a diverse range of plant and animal species. This can be achieved through composting, mulching, and using cover crops to protect and nourish the soil.

Plan for water management: Water is a critical resource for any permaculture system. Designing for resilience includes planning for both water abundance and scarcity. By incorporating strategies such as rainwater harvesting, swales, and drought-tolerant plant species, you can help ensure that your permaculture system remains productive and healthy even during water stress.

Encourage natural pest control: By creating a diverse and interconnected ecosystem, you can encourage natural predators to help control pest populations. This not only reduces the need for chemical interventions but also contributes to the overall resilience of your permaculture system. For example, planting a variety of flowering plants can help to attract beneficial insects while providing habitat for birds and other predators can help to keep pest populations in check.

By incorporating these strategies into your permaculture design, you can create a diverse and resilient ecosystem that can better adapt to environmental changes. As we face increasing challenges from climate change and other environmental pressures, designing for resilience is not only a practical approach but also a necessary one for the long-term success of our permaculture systems.

Integrating Natural Elements and Processes into Your Permaculture System

In order to create a thriving permaculture system, it is essential to integrate natural elements and processes into your design. By working with nature rather than against it, you can harness its power to create a more diverse and resilient ecosystem. This section will explore ways to incorporate natural elements and processes into your permaculture system, ultimately leading to a more sustainable and harmonious environment.

Observe and Mimic Nature

The first step in integrating natural elements and processes into your permaculture system is to carefully observe the natural world around you. By studying the patterns and relationships in nature, you can gain valuable insights into designing your own system. For example, you might notice how certain plants grow well together or how specific animals contribute to the health of an ecosystem. You can create a more diverse and resilient permaculture system by mimicking these natural relationships.

Utilize Natural Energy Sources

One of the fundamental principles of permaculture is using renewable resources and natural energy sources. This can include harnessing the power of the sun, wind, and water to provide energy for your system. By incorporating solar panels, wind turbines, or hydroelectric systems into your design, you can reduce your reliance on non-renewable energy sources and create a more sustainable ecosystem.

Encourage Natural Pest Control

Instead of relying on harmful chemicals to control pests, consider integrating natural pest control methods into your permaculture system. This can include introducing beneficial insects, such as ladybugs and lacewings, which prey on harmful pests. Additionally, planting certain species of plants can help deter pests, as their strong scents or natural toxins can repel unwanted insects. By encouraging natural pest control, you can reduce the need for harmful pesticides and create a healthier ecosystem.

Build Healthy Soil

Healthy soil is the foundation of any successful permaculture system. In order to create a diverse and resilient ecosystem, it is essential to focus on building and maintaining healthy soil. This can be achieved through various methods, such as composting, mulching, and no-till gardening. By nurturing the soil, you can support a diverse array of plant and animal life, ultimately leading to a more resilient ecosystem.

Incorporate Water Management Strategies

Water is a critical component of any permaculture system, and it is essential to manage it effectively. By integrating natural water management strategies, such as rainwater harvesting, swales, and greywater recycling, you can ensure that your system has a consistent and sustainable water supply. These methods not only help conserve water but also contribute to your ecosystem's overall health and resilience.

In conclusion, integrating natural elements and processes into your permaculture system is crucial for creating a diverse and resilient ecosystem. By observing and mimicking nature, utilizing natural energy sources, encouraging natural pest control, building healthy soil, and incorporating water management strategies, you can create a sustainable and harmonious environment that thrives in the face of environmental challenges.

The Lasting Impact of Embracing Diversity and Resilience in Permaculture

In conclusion, the journey towards creating diverse and resilient ecosystems in permaculture is not only a rewarding endeavor but also a crucial step in ensuring the long-term sustainability and success of our agricultural systems. By embracing the principles of diversity and resilience, we can cultivate landscapes that are more adaptable to environmental changes, more productive, and more beneficial to both humans and the natural world.

The importance of fostering diversity in permaculture cannot be overstated. Incorporating various plant and animal species into our systems can create a more balanced and harmonious ecosystem that is less susceptible to pests, diseases, and other threats. This diversity also provides a wealth of resources for humans, from nutritious food to valuable materials and medicines.

Resilience, on the other hand, is the ability of an ecosystem to adapt and thrive in the face of change. By designing our permaculture systems with resilience in mind, we can ensure they will continue flourishing even as the climate and other environmental factors shift. This adaptability is essential for the survival of our agricultural systems and the well-being of the communities that rely on them.

Integrating natural elements and processes into our permaculture systems is another critical aspect of building diverse and resilient ecosystems. By working with nature rather than against it, we can create systems that are more in tune with the natural world and better equipped to handle the challenges that come their way.

As we move forward in our quest to create more sustainable and regenerative agricultural systems, let us remember the importance of embracing diversity and resilience in our permaculture designs. By doing so, we can ensure that our landscapes not only provide for our needs but also contribute to the health and well-being of the planet as a whole. In the end, the lasting impact of diverse and resilient permaculture ecosystems will be felt not only by those who practice it but also by future generations who will inherit a more sustainable and vibrant world.

Chapter Summary

- Diversity and resilience are foundational principles in permaculture, essential for creating sustainable and adaptable ecosystems that can withstand challenges such as climate change, habitat loss, and dwindling resources.
- Plant and animal species are crucial in building ecosystem diversity, contributing to improved soil health, pest and disease control, pollinator attraction, nutrient cycling, and habitat creation.
- Designing for resilience involves embracing diversity, utilizing polycultures, designing for redundancy, fostering

soil health, and planning for water management to create ecosystems that can adapt to environmental changes.
- Integrating natural elements and processes, such as observing and mimicking nature, utilizing natural energy sources, encouraging natural pest control, building healthy soil, and incorporating water management strategies, is essential for creating a diverse and resilient permaculture system.
- Embracing diversity in permaculture leads to more balanced and harmonious ecosystems, providing a wealth of resources for humans and reducing susceptibility to pests, diseases, and other threats.
- Resilience in permaculture ensures that ecosystems can adapt and thrive in the face of change, which is essential for the survival of agricultural systems and the well-being of communities that rely on them.
- Prioritizing diversity and resilience in permaculture designs contributes to the health and well-being of the planet, creating a lasting impact that will benefit both current practitioners and future generations.

7

INTEGRATING LIVESTOCK AND WILDLIFE

A whimsical farm scene where livestock and wildlife animals are having a friendly picnic together, with cows, deer, pigs, rabbits, and birds sharing food and enjoying each other's company under a large, shady tree.

In the world of permaculture, the integration of livestock and wildlife is a vital component in creating a thriving, sustainable ecosystem. This harmonious relationship between domesticated animals and their wild counterparts is not only fascinating but also essential for the long-term success of any permaculture system. By understanding and nurturing this symbiotic relationship, we can cultivate a productive and resilient environment that benefits all living beings within it.

At the heart of permaculture lies the principle of **working with nature** rather than against it. This means recognizing the inherent value of every living organism and the role it plays in the ecosystem. Livestock and wildlife are no exception. By integrating these two elements, we can tap into their unique strengths and abilities, creating a balanced and diverse landscape that supports a wide range of species.

Livestock can provide numerous benefits to a permaculture system, such as producing food, improving soil fertility, and controlling pests. At the same time, wildlife contributes to the overall health of the ecosystem by pollinating plants, dispersing seeds, and maintaining a balance of predator and prey populations. When these two groups coexist and interact, they form a mutually beneficial relationship that enhances the overall productivity and resilience of the permaculture system.

This chapter will explore how livestock and wildlife can be integrated into a permaculture design, from choosing the right animals for your system to creating habitats that encourage wildlife to thrive. We will also discuss how to manage interactions between livestock and wildlife to ensure that both groups benefit from their coexistence and address potential challenges and conflicts that may arise.

By understanding the symbiotic relationship between livestock and wildlife, we can create a permaculture system that not only produces food and resources for human consumption but also supports a diverse and thriving ecosystem. This holistic approach to land management is the key to building a sustainable future for ourselves and the countless species with whom we share this planet.

Choosing the Right Livestock for Your Permaculture System

Selecting the appropriate livestock for your permaculture system ensures a harmonious and productive integration of animals and plants. The right livestock can provide numerous benefits, such as nutrient cycling, pest control, and food production. This section will discuss the factors to consider when choosing livestock and offer some examples of animals that can be successfully incorporated into a permaculture system.

Assessing Your Needs and Goals: Before diving into the world of livestock, it's essential to assess your specific needs and goals for your permaculture system. Are you looking to produce food for your family, or are you more focused on creating a sustainable ecosystem? Your objectives will help guide your decision-making when selecting the right animals for your system.

Climate and Environment: The climate and environment of your permaculture site play a significant role in determining which livestock will thrive. Some animals are better suited to colder climates, while others prefer warmer temperatures. Additionally, the availability of water, vegetation, and shelter will impact the types of animals that can be successfully integrated into your system.

Size and Space Requirements: The size of your permaculture site and the space available for livestock will also influence your choices. Smaller animals, such as chickens and rabbits, require less space and can be easily accommodated in most permaculture systems. However, larger animals, like cows and goats, need more room to roam and graze, so it's essential to ensure that you have adequate space to meet their needs.

Skills and Experience: Your level of experience and comfort with handling livestock should also be considered when selecting animals for your permaculture system. If you are new to raising livestock, it's wise to start with low-maintenance animals, such as chickens or ducks, before moving on to more challenging species.

Examples of Livestock for Permaculture Systems

Here are some examples of livestock that can be successfully integrated into a permaculture system:

- **Chickens:** These versatile birds provide eggs, meat, and natural pest control. They can also help with composting by turning and aerating the soil.
- **Ducks:** Ducks are excellent foragers and can help control pests like slugs and snails. They also produce eggs and meat.
- **Goats:** These hardy animals can provide milk, meat, and fiber, while also helping to control weeds and brush.
- **Sheep:** Sheep can offer meat, wool, and milk, and they are efficient grazers that can help maintain pasture health.
- **Pigs:** Pigs are natural rototillers, turning over soil and helping to break up compacted ground. They can also provide meat and help with composting.
- **Bees:** These essential pollinators can increase your garden's productivity and provide honey.

In conclusion, choosing the right livestock for your permaculture system involves considering your needs, goals, climate, available space, and experience. By carefully selecting the appropriate animals, you can create a thriving ecosystem that benefits both livestock and wildlife.

Creating Habitats for Wildlife to Thrive

In a thriving permaculture system, wildlife plays a crucial role in maintaining balance and harmony. By creating habitats that cater to the needs of various species, you can encourage a diverse range of wildlife to flourish alongside your livestock. This section will explore the different ways you can create and enhance habitats for wildlife to thrive in your permaculture system.

First and foremost, it's essential to understand the specific needs of the wildlife species you wish to attract. Each species has unique requirements for food, water, shelter, and breeding sites. Providing these essentials allows you to create a welcoming environment that encourages wildlife to make your permaculture system their home.

One effective way to create habitats for wildlife is by planting a diverse range of **native plants.** Native plants are well-adapted to the local climate and soil conditions, making them a perfect choice for supporting local wildlife. These plants provide food, shelter, and nesting sites for various species, including pollinators, birds, and small mammals. Additionally, native plants often require less maintenance and water, making them a sustainable choice for your permaculture system.

Water sources are another critical component of wildlife habitats. By incorporating ponds, streams, or even small birdbaths into your permaculture design, you can attract a variety of species that depend on water for survival. Ensure that water sources are clean and accessible. Consider adding rocks or logs to provide additional shelter and basking spots for amphibians and reptiles.

Another essential aspect of creating habitats for wildlife to thrive is **shelter.** Providing a variety of shelter options, such as brush piles, rock piles, and nesting boxes, can accommodate the needs of different species. These shelters offer protection from predators and harsh weather conditions, as well as nesting and breeding sites.

In addition to providing food, water, and shelter, it's crucial to consider the connectivity of your permaculture system. Wildlife corridors like hedgerows and tree lines allow animals to move safely between different habitats. These corridors are essential for maintaining genetic diversity and ensuring the long-term survival of various species.

Finally, practicing responsible **land management** is essential to create habitats for wildlife to thrive. This includes minimizing the use of pesticides and herbicides, which can harm wildlife and your livestock. Instead, opt for natural pest control methods, such as introducing beneficial insects and practicing crop rotation.

In conclusion, creating habitats for wildlife to thrive is an integral part of a successful permaculture system. By providing food, water, shelter, and connectivity essentials, you can encourage diverse species to coexist harmoniously with your livestock. This benefits the wildlife and contributes to the overall health and productivity of your permaculture system.

Managing Livestock and Wildlife Interactions for Mutual Benefit

In a well-designed permaculture system, livestock and wildlife can coexist harmoniously, each benefiting from the presence of the other. This section will explore various strategies for managing livestock and wildlife interactions to ensure a mutually beneficial relationship.

Encouraging Natural Pest Control

One of the primary benefits of integrating livestock and wildlife in a permaculture system is the natural pest control that can be achieved. Many wild animals, such as birds, frogs, and insects, feed on common pests that can damage crops and harm livestock. Creating habitats that attract these beneficial creatures can reduce the need for chemical pesticides and promote a healthier ecosystem.

To **encourage natural pest control**, consider the following strategies:

- Plant a variety of native plants that provide food and shelter for beneficial insects and birds.
- Install birdhouses, bat boxes, and other nesting sites to attract insect-eating species.
- Create small ponds or wetland areas to support amphibians and other aquatic wildlife.
- Avoid using chemical pesticides, which can harm beneficial species as well as pests.

Grazing Management for Soil Health

Grazing livestock, such as cows, sheep, and goats, can be crucial in maintaining soil health and fertility. Their grazing habits help to control weeds, aerate the soil, and recycle nutrients through their manure. However, it's essential to manage grazing patterns to prevent overgrazing and soil compaction.

Rotational grazing is an effective strategy for managing livestock and promoting soil health. This involves dividing your land into smaller paddocks and moving your livestock between them on a regular basis. This allows the vegetation in each paddock to recover and prevents overgrazing. Additionally, rotational grazing can help to distribute manure more evenly, providing a natural source of fertilizer for your crops.

Promoting Pollination and Biodiversity

Wildlife, particularly pollinators such as bees, butterflies, and hummingbirds, play a vital role in the success of your permaculture system. By providing food, water, and shelter for these species, you can encourage their presence and increase the productivity of your crops.

To promote pollination and biodiversity, consider the following strategies:

- Plant diverse flowering plants, including native species that bloom at different times throughout the year.
- Provide water sources, such as shallow dishes or birdbaths, for pollinators to drink.
- Leave some areas of your land undisturbed, allowing for natural habitats to form and support a variety of wildlife species.

Monitoring and Adapting to Changing Conditions

Managing livestock and wildlife interactions requires ongoing observation and adaptation. Regularly monitor the health of your animals, the condition of your land, and the presence of wildlife to identify any potential issues or conflicts. By staying attuned to the needs of your permaculture system, you can make adjustments as needed to maintain a harmonious and productive environment.

In conclusion, integrating livestock and wildlife into your permaculture system can lead to numerous long-term benefits, including natural pest control, improved soil health, and increased crop productivity. By carefully managing these interactions and creating a supportive environment for both livestock and wildlife, you can cultivate a thriving and sustainable ecosystem.

Addressing Challenges and Conflicts Between Livestock and Wildlife

In an ideal permaculture system, livestock and wildlife coexist harmoniously, each contributing to the overall health and productivity of the ecosystem. However, it is essential to recognize that challenges and conflicts may arise between these two groups. In this section, we will discuss some common issues and provide practical solutions to ensure a balanced and thriving environment for both livestock and wildlife.

One of the most common challenges in integrating livestock and wildlife is predation. Predators such as foxes, coyotes, and birds of prey can threaten your livestock, particularly smaller animals like chickens and rabbits. To protect your livestock from predators, consider implementing the following strategies:

- Provide secure and well-constructed animal housing, ensuring that all potential entry points are covered and reinforced.
- Employ guardian animals, such as dogs, llamas, or donkeys, to deter predators and protect your livestock.
- Install fencing or other barriers to keep predators out of your property. Electric fencing can be particularly effective in deterring larger predators.
- Encourage the presence of natural predators of smaller predators, such as owls and hawks, by providing nesting boxes and perches.

Another challenge in integrating livestock and wildlife is competition for resources like food and water. To minimize competition and ensure that both groups have access to the resources they need, consider the following strategies:

- Plant a diverse range of forage species to provide ample food for both livestock and wildlife. This can include grasses, legumes, and other plants that provide both nutrition and habitat for various species.
- Establish separate feeding and watering stations for livestock and wildlife, ensuring each group has access to clean and uncontaminated resources.
- Rotate your livestock through different areas of your property, allowing forage to regrow and preventing overgrazing.

Disease transmission between livestock and wildlife is another concern that must be addressed when integrating these two groups. To minimize the risk of disease transmission, consider the following strategies:

- Maintain a healthy and diverse ecosystem, which can help prevent disease spread by supporting a robust population of beneficial insects, birds, and other wildlife.
- Implement biosecurity measures, such as quarantining new animals and regularly monitoring the health of your livestock.
- Encourage natural pest control by providing a habitat for beneficial insects and predators, such as ladybugs, lacewings, and birds.

Lastly, it is crucial to address any legal or regulatory challenges that may arise when integrating livestock and wildlife. Be sure to research and comply with any local, state, or federal regulations regarding the management of wildlife and livestock on your property.

In conclusion, addressing challenges and conflicts between livestock and wildlife is crucial to creating a successful permaculture system. By implementing practical solutions and maintaining a proactive approach to management, you can ensure a harmonious and productive environment for both groups, ultimately contributing to the long-term sustainability and resilience of your permaculture system.

The Long-term Benefits of Integrating Livestock and Wildlife in Permaculture

In conclusion, integrating livestock and wildlife in permaculture systems offers many long-term benefits that contribute to our environment's overall health and sustainability. By fostering a symbiotic relationship between these two essential components, we can create a thriving ecosystem that benefits not only the plants and animals within it but also the people who depend on it for their livelihood and well-being.

One of the most significant long-term benefits of integrating livestock and wildlife in permaculture is **biodiversity enhancement**. By providing a variety of habitats and resources for different species, we can support a rich and diverse community of plants, animals, and microorganisms. This biodiversity, in turn, contributes to the resilience of the ecosystem, enabling it to better withstand disturbances and adapt to changing conditions.

Another long-term benefit is the improvement of soil health and fertility. Livestock and wildlife play a crucial role in cycling nutrients and organic matter through the ecosystem, helping to build and maintain healthy, fertile soils. This not only supports the growth of more robust and productive plants but also helps to sequester carbon and mitigate the impacts of climate change.

Integrating livestock and wildlife in permaculture systems also contributes to pest management and disease control. Encouraging diverse species can create a more balanced ecosystem where natural predators help keep pest populations in check. This reduces the need for harmful chemical pesticides and promotes a healthier environment for all.

Moreover, the integration of livestock and wildlife can provide valuable economic benefits for farmers and landowners. They can build a more resilient and sustainable business model by diversifying their operations and incorporating multiple income streams, such as meat, dairy, eggs, and ecotourism. This, in turn, supports local economies and contributes to food security in the community.

Lastly, integrating livestock and wildlife in permaculture systems fosters a deeper connection between humans and the natural world. By working in harmony with nature, we can cultivate a greater appreciation for the intricate web of life that sustains us and develop a more profound sense of stewardship for the land and its inhabitants.

In summary, the long-term benefits of integrating livestock and wildlife in permaculture are vast and far-reaching. By embracing this holistic approach, we can create healthier, more resilient ecosystems that support a diverse array of species, improve soil health, and contribute to the well-being of both humans and the environment. As we continue to face the challenges of a rapidly changing world, the principles of permaculture offer a guiding light towards a more sustainable and harmonious future.

Chapter Summary

- Integrating livestock and wildlife in permaculture systems is essential for creating a sustainable and thriving ecosystem, as it fosters a symbiotic relationship between domesticated animals and their wild counterparts.
- Choosing the right livestock for your permaculture system involves considering your needs and goals, climate, available space, and experience with handling animals.
- Creating habitats for wildlife to thrive is crucial for maintaining a balanced and diverse landscape, which can be

achieved by providing food, water, shelter, and connectivity for various species.

- Managing livestock and wildlife interactions for mutual benefit includes encouraging natural pest control, practicing rotational grazing for soil health, promoting pollination and biodiversity, and monitoring and adapting to changing conditions.
- Addressing challenges and conflicts between livestock and wildlife, such as predation, competition for resources, and disease transmission, is vital for maintaining a harmonious and productive environment.
- Complying with legal and regulatory requirements is essential when integrating livestock and wildlife in permaculture systems.
- The long-term benefits of integrating livestock and wildlife in permaculture include enhanced biodiversity, improved soil health and fertility, natural pest management and disease control, economic benefits for farmers and landowners, and fostering a deeper connection between humans and the natural world.
- Embracing the principles of permaculture and integrating livestock and wildlife can help create healthier, more resilient ecosystems that support a diverse array of species and contribute to the well-being of both humans and the environment.

8

NATURAL PEST CONTROL AND MANAGEMENT

Ladybugs and lacewings teaming up to protect a vibrant garden from aphids and other pests.

In the world of permaculture, the concept of working with nature rather than against it is a fundamental principle. As we strive to create sustainable and self-sufficient ecosystems, one of the most significant challenges we face is **managing pests** in a way that aligns with our commitment to environmental stewardship. Enter the realm of natural pest control and management, a holistic approach that seeks to harness the power of nature to keep pests in check and maintain the balance of our gardens and landscapes.

At first glance, natural pest control and management might seem counterintuitive. After all, aren't pests the very thing we're trying to eliminate from our gardens? However, the key is understanding that not all insects and organisms are harmful. In fact, many play crucial roles in maintaining the health and vitality of our ecosystems. By focusing on promoting the presence of these beneficial creatures and employing organic techniques to deter pests, we can create a thriving permaculture garden that is both productive and ecologically sound.

This chapter will explore the fascinating world of natural pest control and management, exploring the principles and practices that can help you transform your garden into a harmonious and balanced ecosystem. We'll begin by examining the importance of understanding the ecosystem and the role of beneficial insects in maintaining equilibrium.

Next, we'll discuss various organic pest control techniques and strategies that can be employed to minimize the impact of harmful pests on your garden. We'll also cover the concept of companion planting, a powerful tool for managing pests through strategic plant selection and placement. Finally, we'll discuss the importance of monitoring and maintaining a healthy permaculture garden, ensuring that your efforts in natural pest control and management continue to yield positive results.

As we embark on this journey, remember that the goal is not to wage war on pests but to embrace nature's wisdom and work in harmony with the intricate web of life that sustains us all. By adopting a natural approach to pest control and management, you'll contribute to your garden's health and the planet's well-being as a whole. So let's get started, and discover the incredible potential of natural pest control and management in the world of practical permaculture.

Understanding the Ecosystem and Beneficial Insects

In order to effectively implement natural pest control and management in your permaculture garden, it is essential to first understand the ecosystem and the role of beneficial insects. By gaining a deeper comprehension of the intricate relationships between plants, insects, and other organisms, you can harness nature's wisdom to create a thriving, balanced garden that minimizes the need for synthetic pesticides and other harmful interventions.

The ecosystem within your garden is a **complex web of interactions** between various living organisms, including plants, insects, birds, and microorganisms. Each of these components plays a vital role in maintaining the overall health and balance of the garden. In particular, beneficial insects are crucial allies in your quest for natural pest control and management.

Beneficial insects can be divided into two main categories: **predators** and **pollinators.** Predatory insects, such as ladybugs, lacewings, and predatory mites, help control pest populations by feeding on common garden pests like aphids, mites, and caterpillars. These natural predators are essential to any healthy ecosystem and can significantly reduce the need for chemical pesticides.

Pollinators, on the other hand, are essential for the reproduction and growth of many plants. Insects like bees, butterflies, and certain beetles play a crucial role in transferring pollen from one flower to another, allowing plants to produce fruits and seeds. You can ensure a bountiful harvest and a vibrant, thriving garden by attracting and supporting a diverse range of pollinators.

To effectively utilize beneficial insects in your permaculture garden, it's important to create a **welcoming environment** that supports their needs. This can be achieved by:

- Planting diverse flowering plants that provide nectar and pollen for pollinators throughout the growing season.
- Providing habitat for beneficial insects, such as shrubs, trees, and ground cover, where they can find shelter, reproduce, and overwinter.
- Avoid synthetic pesticides, which can harm pests and beneficial insects. Instead, opt for organic pest control methods that target specific pests while minimizing harm to beneficial insects.
- Introducing beneficial insects to your garden, either by purchasing them from a reputable supplier or by attracting them through specific plants and habitat features.

By understanding the ecosystem and the role of beneficial insects, you can create a permaculture garden that works in harmony with nature rather than against it. This approach not only leads to a healthier, more productive garden but also contributes to the overall well-being of our planet. The following sections will explore various organic pest control techniques and strategies to help you maintain a balanced and thriving permaculture garden.

Organic Pest Control Techniques and Strategies

In this section, we will delve into the world of organic pest control techniques and strategies that are both effective and environmentally friendly. These methods not only help maintain the balance of your permaculture garden but also contribute to the overall health of the ecosystem. By adopting these techniques, you can manage pests without resorting to harmful chemical pesticides, ensuring a safe and thriving environment for all living beings in your garden.

Biological Control: One of the most effective organic pest control methods is introducing natural predators and parasites into your garden. These beneficial insects, such as ladybugs, lacewings, and parasitic wasps, can help keep pest populations in check by preying on them or laying their eggs inside the pests. To attract these helpful insects, plant a variety of flowering plants that provide nectar and pollen, and consider purchasing and releasing them in your garden if they are not naturally present.

Physical Barriers: Erecting physical barriers can effectively prevent pests from accessing your plants. Examples of such barriers include floating row covers, which can be placed over your plants to protect them from flying insects, and copper tape, which can be wrapped around the base of plants to deter slugs and snails. Additionally, using netting or fencing can help keep larger pests, such as rabbits and deer, at bay.

Cultural Practices: Adopting certain cultural practices can help reduce the likelihood of pest infestations. For instance, practicing crop rotation can disrupt the life cycle of pests that are specific to certain plants, while proper sanitation, like removing dead leaves and debris, can eliminate potential hiding places for pests. Moreover, maintaining healthy soil and providing your plants with the appropriate nutrients can help them better withstand pest attacks.

Mechanical Control: This involves physically removing or trapping pests from your plants. Handpicking pests, such as caterpillars and beetles, can be an effective method for small gardens. For larger infestations, you can use traps, such as yellow sticky traps to catch flying insects or pheromone traps to attract and capture specific pests.

Botanical and Mineral-Based Pesticides: In cases where other organic pest control methods are insufficient, you can opt for botanical and mineral-based pesticides. These are derived from natural sources and are generally less harmful to the environment than synthetic chemical pesticides. Examples include neem oil, which can control a variety of pests, and diatomaceous earth, which can be used to control crawling insects. However, it is essential to use these products judiciously and follow the label instructions to minimize any potential harm to beneficial insects.

By implementing these organic pest control techniques and strategies, you can create a balanced and healthy permaculture garden that can thrive without the need for harmful chemicals. In the next section, we will explore the concept of companion planting and how it can further enhance your garden's pest management efforts.

Implementing Companion Planting for Pest Management

In the world of permaculture, **companion planting** is a time-tested and highly effective strategy for managing pests naturally and sustainably. By harnessing the power of plant relationships and synergies, you can create a thriving garden ecosystem less susceptible to pest infestations and more resilient in the face of challenges. In this section, we will explore the principles of companion planting, discuss some popular plant combinations, and offer practical tips for implementing this strategy in your own permaculture garden.

Companion planting is based on the idea that certain plants can benefit each other when grown in close proximity. This can happen in several ways, such as attracting beneficial insects, repelling pests, or providing support and nutrients to neighboring plants. The key to successful companion planting is understanding the specific needs and characteristics of each plant species and selecting companions that complement and enhance their growth.

One of the most common uses of companion planting in permaculture is to attract beneficial insects that prey on pests. For example, planting flowers like marigolds, calendula, and yarrow near your vegetable crops can draw in ladybugs, lacewings, and parasitic wasps, which are natural predators of aphids, caterpillars, and other common garden pests. By providing a habitat for these helpful insects, you can reduce the need for chemical pesticides and promote a more balanced and diverse ecosystem.

Another critical aspect of companion planting is choosing plants that repel or deter pests. Some plants emit strong odors or contain unappealing or toxic compounds to certain insects, making them effective natural pest control agents. For instance, planting garlic or onions near lettuce can help keep aphids at bay, while basil planted alongside tomatoes can deter tomato hornworms. Similarly, intercropping your garden with strong-smelling herbs like mint, rosemary, and thyme can confuse and repel a wide range of pests, protecting your valuable crops from damage.

In addition to these direct pest control benefits, companion planting can also improve your garden's overall health and productivity by fostering beneficial plant relationships. For example, planting nitrogen-fixing legumes like beans and peas near heavy feeders like corn and squash can help improve soil fertility and reduce the need for synthetic fertilizers. Similarly, planting tall, sturdy plants like sunflowers or corn can provide natural support and shade for vining or low-growing plants like beans and cucumbers, creating a more efficient and harmonious garden layout.

To implement companion planting in your permaculture garden, start by researching the specific needs and characteristics of your chosen plants and identifying potential companion species that can provide mutual benefits. Then, plan your garden layout with these relationships in mind, grouping compatible plants together and considering factors like sunlight, water, and nutrient requirements. Be prepared to experiment and adjust your plantings as you observe the results and learn from your garden's unique ecosystem.

In conclusion, companion planting is a powerful and practical tool for managing pests and promoting a healthy, productive permaculture garden. By understanding and embracing the natural relationships between plants, you can harness nature's wisdom to create a more resilient and sustainable gardening system. So, go ahead and explore the world of companion planting, and watch your garden flourish in harmony with the environment.

Monitoring and Maintaining a Healthy Permaculture Garden

A thriving permaculture garden is a delicate balance of various elements working in harmony with nature. It is crucial to monitor and maintain your garden's health regularly to ensure its success. This section will guide you through the essential steps to keep your permaculture garden flourishing and effectively manage pests using natural methods.

First and foremost, **observation is vital.** Regularly walk through your garden and observe the plants, insects, and overall health of the ecosystem. This will help you identify any potential issues before they escalate, such as signs of pest infestations, diseases, or nutrient deficiencies. Make a habit of inspecting your garden at least once a week or more frequently during critical growth periods or when introducing new plants or techniques.

One essential aspect of maintaining a healthy permaculture garden is ensuring proper soil health. Healthy soil is the foundation of a thriving ecosystem, as it provides the necessary nutrients and support for plants to grow. Regularly test your soil to monitor its pH levels, nutrient content, and structure. Amend the soil with organic matter, such as compost or well-rotted manure, to maintain its fertility and structure. Additionally, practice crop rotation and cover cropping to prevent nutrient depletion and soil-borne diseases.

Another crucial factor in maintaining a healthy garden is proper watering. Overwatering or underwatering can lead to stressed plants, making them more susceptible to pests and diseases. Observe the moisture levels in your soil and adjust your watering schedule accordingly. Utilize drip irrigation or rainwater harvesting techniques to conserve water and ensure that your plants receive the appropriate amount of moisture.

Pruning and trimming plants is also an essential part of garden maintenance. Regularly remove dead or diseased plant material to prevent the spread of diseases and pests. Pruning also promotes air circulation and sunlight penetration, which are vital for plant health and growth. Be sure to sanitize your pruning tools between uses to avoid spreading diseases.

Attracting and supporting beneficial insects is a cornerstone of natural pest control in a permaculture garden. Encourage these helpful creatures by providing diverse plants, including flowering species that produce nectar and pollen. Additionally, consider installing insect hotels or nesting boxes to provide shelter and breeding sites for beneficial insects.

Finally, be prepared to intervene when necessary. Despite your best efforts, some pests may still find their way into your garden. In such cases, act promptly and use organic pest control methods, such as introducing predatory insects, using insecticidal soaps, or applying natural repellents like neem oil. Always opt for the least harmful method first, and do not escalate unless it is required.

In conclusion, monitoring and maintaining a healthy permaculture garden is an ongoing process that requires patience, observation, and a deep understanding of the natural ecosystem. By embracing nature's wisdom and working in harmony with it, you can create a thriving garden that effectively manages pests and contributes to a sustainable future.

Embracing Nature's Wisdom in Pest Control and Management

As we reach the end of this enlightening journey through natural pest control and management, it is essential to reflect on the wisdom that nature has to offer. Throughout this chapter, we have explored the importance of understanding the ecosystem, the role of beneficial insects, organic pest control techniques, and the power of companion planting. All these elements work together to create a harmonious and healthy permaculture garden that thrives without the need for harmful chemicals.

Embracing nature's wisdom in pest control and management means recognizing that our gardens are not isolated entities but are part of a larger ecosystem. By working with nature rather than against it, we can create an environment where pests are kept in check by natural predators and plants that support each other's growth. This approach not only benefits our plants but also contributes to the overall health of our environment and the planet.

One of the most important aspects of natural pest control and management is the **willingness to learn from nature.** Observing the interactions between plants, insects, and other organisms in our gardens can provide valuable insights into how we can best support the health and productivity of our plants. By adopting a holistic approach to pest management, we can minimize the need for intervention and create a self-sustaining garden that thrives with minimal input from us.

In conclusion, the principles of practical permaculture encourage us to work with nature to create a sustainable and productive garden. By understanding the ecosystem, harnessing the power of beneficial insects, implementing organic pest control techniques, and utilizing companion planting, we can create a garden that is not only beautiful but also resilient and self-sustaining. Embracing nature's wisdom in pest control and management is a practical approach to gardening and a testament to our commitment to preserving the environment for future generations.

As you continue your journey in permaculture, remember to always be open to learning from nature, observing the intricate relationships between organisms, and seeking innovative ways to create a harmonious and thriving garden. With patience, dedication, and deep respect for the natural world, you will undoubtedly reap the rewards of a flourishing permaculture garden that is both productive and sustainable.

Chapter Summary

- Natural pest control and management is a holistic approach that works with nature to maintain a balanced and healthy garden ecosystem, minimizing the need for synthetic pesticides and other harmful interventions.
- Understanding the ecosystem and the role of beneficial insects, such as predators and pollinators, is crucial for effective natural pest control and management in a permaculture garden.
- Organic pest control techniques and strategies, including biological control, physical barriers, cultural practices, mechanical control, and botanical and mineral-based pesticides, can help manage pests without resorting to harmful chemicals.

- Companion planting is a powerful tool for managing pests through strategic plant selection and placement, harnessing the power of plant relationships and synergies to create a thriving garden ecosystem.
- Regular monitoring and maintaining a healthy permaculture garden is essential for its success, involving observation, proper soil health, appropriate watering, pruning, and supporting beneficial insects.
- Attracting and supporting beneficial insects is a cornerstone of natural pest control in a permaculture garden, achieved by providing a diverse range of plants and installing insect hotels or nesting boxes.
- Embracing nature's wisdom in pest control and management means recognizing that gardens are part of a larger ecosystem and working with nature to create an environment where pests are kept in check by natural predators and supportive plants.
- By adopting a holistic approach to pest management and learning from nature, we can create a self-sustaining garden that thrives with minimal input, contributing to the overall health of our environment and the planet.

9

PERMACULTURE IN URBAN AND SMALL SPACES

An enchanting urban rooftop garden, incorporating elements of permaculture such as vertical planting, rainwater harvesting, and companion planting, to create a lush green oasis amidst the concrete jungle.

Permaculture is a holistic approach to sustainable living that integrates ecological design, resource conservation, and self-reliance. While it may be easy to envision permaculture principles being applied to sprawling rural landscapes, embracing these concepts in urban and small spaces is equally essential and possible. In this chapter, we will explore the myriad ways in which permaculture can be adapted to fit the unique challenges and opportunities presented by limited spaces.

The world is becoming increasingly urbanized, with more than half the global population residing in cities. This trend is expected to continue, making it essential for *urban dwellers* to find innovative ways to grow food, conserve resources, and build resilient communities in the face of climate change and other challenges. Permaculture offers a powerful framework for achieving these goals, even in the most cramped and concrete-laden environments.

Adopting a mindset of creativity, flexibility, and resourcefulness in embracing permaculture in limited spaces is crucial. This means looking beyond traditional gardening methods and embracing innovative techniques such as vertical gardening, multi-functional elements, and urban food forests. It also involves recognizing the value of community connections and the potential for integrating small-scale livestock into urban ecosystems.

This chapter will delve into these topics and provide practical guidance for implementing permaculture principles in urban and small spaces. By doing so, we aim to empower readers to transform their balconies, rooftops, and tiny backyards into thriving, resilient ecosystems that contribute to the health and well-being of both people and the planet.

In the following sections, we will explore the diverse ways in which permaculture can be adapted to urban and small spaces, from maximizing space through vertical gardening and multi-functional elements to creating edible landscapes and fostering community connections. We will also discuss the potential for integrating small-scale livestock into urban permaculture systems, highlighting the benefits and challenges of doing so. By the end of this chapter, readers will be equipped with the knowledge and inspiration needed to embrace permaculture in their own limited spaces, transforming them into thriving, sustainable ecosystems.

Maximizing Space: Vertical Gardening and Multi-Functional Elements

In the world of permaculture, creativity knows no bounds, especially when it comes to maximizing space in urban and small-scale settings. With limited room to work with, it's essential to think outside the box and utilize every inch of available space. In this section, we'll explore two critical strategies for making the most of your urban or small-space permaculture garden: **vertical gardening** and **multi-functional elements.**

Vertical Gardening

When horizontal space is scarce, it's time to look up and embrace the concept of **vertical gardening.** This innovative approach allows you to grow plants on walls, trellises, and other vertical structures, effectively multiplying your garden's square footage without expanding its footprint. Vertical gardening saves space and offers numerous benefits, such as improved air circulation, easier pest control, and increased exposure to sunlight for your plants.

To get started with vertical gardening, consider the following techniques:

Trellises and Climbing Plants: Install trellises or other support structures for climbing plants like beans, peas, cucumbers, and even some varieties of tomatoes. These plants will naturally grow upwards, making the most of your vertical space.

Wall Planters: Attach planters to walls, fences, or other vertical surfaces to create a living wall of herbs, flowers, or even small fruiting plants like strawberries. This not only maximizes space but also adds visual interest to your garden.

Pallet Gardens: Repurpose wooden pallets by filling them with soil and planting a variety of shallow-rooted plants in the gaps between the slats. Stand the pallets upright against a wall or fence for an instant vertical garden.

Hanging Baskets: Suspend hanging baskets from balconies, eaves, or other overhead structures to grow trailing plants like nasturtiums, cherry tomatoes, or even small fruit trees.

Multi-Functional Elements

Another key strategy for maximizing space in your urban or small-space permaculture garden is to incorporate **multi-functional elements.** These are features that serve more than one purpose, effectively increasing the efficiency and productivity of your garden without taking up additional space.

Here are some examples of multi-functional elements you can incorporate into your garden:

1. **Edible Landscaping:** Replace purely ornamental plants with edible varieties that provide both beauty and sustenance. For example, plant fruit trees or berry bushes instead of non-fruiting ornamentals or use edible ground covers like creeping thyme or oregano instead of traditional lawn grasses.

2. **Living Fences:** Instead of installing a standard wooden or metal fence, consider planting a hedge of fruiting shrubs or trees, like raspberries or hazelnuts. This not only provides a natural barrier but also offers food and habitat for wildlife.

3. **Rainwater Harvesting:** Collect rainwater from your roof or other surfaces using gutters, downspouts, and rain barrels. This water can then be used to irrigate your garden, reducing your reliance on municipal water supplies and lowering your water bill.

4. **Composting Systems:** Turn kitchen scraps and yard waste into valuable fertilizer by setting up a composting system in your garden. This not only reduces the amount of waste sent to landfills but also provides a steady supply of nutrient-rich compost for your plants.

You can make the most of every available space by embracing vertical gardening and incorporating multi-functional elements into your urban or small-space permaculture garden. In doing so, you'll create a thriving, productive oasis that not only nourishes your body and soul but also contributes to the health and resilience of your local ecosystem.

Urban Food Forests: Creating Edible Landscapes in the City

As cities continue to grow and expand, the need for sustainable and efficient food production becomes increasingly important. Urban food forests offer a unique and innovative solution to this challenge, transforming underutilized spaces into thriving, edible landscapes. In this section, we will explore the concept of urban food forests, their benefits, and how you can create your own edible oasis in the heart of the city.

At its core, an urban food forest is a carefully designed ecosystem that **mimics the structure and function of a natural forest** while providing a diverse array of edible plants. By integrating fruit and nut trees, berry bushes, perennial vegetables, and herbs, these landscapes not only produce food but also provide habitat for wildlife, improve air and water quality, and create a sense of community among city dwellers.

One of the most significant benefits of urban food forests is their ability to produce a high yield of diverse crops in a relatively small area. By utilizing vertical space through the planting of trees, shrubs, and groundcovers, these gardens can produce an abundance of food without the need for large tracts of land. This makes them an ideal solution for urban environments where space is often limited.

Creating an urban food forest begins with careful planning and design. Start by assessing the available space, sunlight, and soil conditions to determine which plants will thrive in your location. Next, consider the various layers of a forest ecosystem – the canopy, understory, shrub layer, herbaceous layer, and ground cover – and select plants that will fill each of these roles. Doing so will create a diverse and resilient system that can withstand pests, diseases, and changing environmental conditions.

When selecting plants for your urban food forest, prioritize native species and those that are well-adapted to your local climate. This will ensure that your garden is low-maintenance and supports local biodiversity. Additionally, choose plants that offer multiple functions, such as nitrogen-fixing legumes that improve soil fertility or fruit trees that provide shade and habitat for birds.

Once your urban food forest is established, it will require minimal intervention to maintain its productivity. Regular pruning, mulching, and occasional watering may be necessary, but the self-sustaining nature of these ecosystems means that they will largely take care of themselves. As the food forest matures, it will become a lush, abundant space that provides food, beauty, and tranquility in the heart of the city.

In conclusion, urban food forests represent a powerful and transformative approach to urban agriculture and permaculture. By harnessing the principles of natural ecosystems and applying them to our city landscapes, we can create abundant, resilient, and beautiful spaces that nourish both people and the planet. So, whether you have a small backyard or a vacant lot, consider the potential of an urban food forest to transform your space into a thriving, edible landscape.

Community Connections: Building Resilience through Local Networks

In the realm of urban and small-space permaculture, the power of community connections cannot be overstated. As we strive to create sustainable, resilient, and vibrant ecosystems in our limited spaces, we must recognize that we are not isolated entities. Instead, we are part of a larger network of individuals, families, and organizations, all of whom can contribute to the success of our permaculture endeavors. In this section, we will explore the importance of building and nurturing local networks to enhance the resilience of our urban and small-space permaculture projects.

The concept of community connections in permaculture is rooted in the understanding that **we are stronger when we work together.** By pooling our resources, knowledge, and skills, we can create more robust and adaptable systems that are better equipped to withstand the challenges of urban living and the uncertainties of our changing world. In practical terms, this means engaging with our neighbors, local organizations, and fellow permaculture enthusiasts to share ideas, collaborate on projects, and support one another in our journey toward a more sustainable future.

One of the most effective ways to build community connections in urban and small-space permaculture is by **establishing local gardening and permaculture groups.** These groups can serve as a platform for individuals to come together to learn from one another, share resources, and collaborate on projects that benefit the entire community. Participating in such groups gives you access to a wealth of knowledge and experience to help you overcome the unique challenges of practicing permaculture in limited spaces.

Another powerful tool for fostering community connections is the **organization of local events**, such as workshops, skill-sharing sessions, and garden tours. These events not only provide valuable learning opportunities but also help to strengthen the bonds between community members and create a sense of shared purpose. Additionally, they can serve as a means of raising awareness about permaculture and its potential benefits for urban and small-space dwellers.

In addition to these formalized groups and events, it is essential to cultivate informal connections with our neighbors and fellow community members. Simple acts such as sharing surplus produce, offering assistance with garden tasks, or exchanging seeds and plant cuttings can go a long way in building trust and fostering a sense of camaraderie.

These informal connections can also lead to more significant collaborations, such as establishing community gardens, creating shared composting systems, or developing neighborhood-wide initiatives to promote sustainable living.

In conclusion, community connections are vital to urban and small-space permaculture. By building and nurturing local networks, we can tap into our community's collective wisdom, resources, and skills, making our permaculture projects more resilient and successful. Furthermore, these connections help to create a sense of belonging and shared purpose, fostering a culture of sustainability and cooperation that extends far beyond the boundaries of our individual gardens and homes.

Small-Scale Livestock: Integrating Animals into Urban Permaculture

When we think of permaculture, we often envision sprawling gardens and vast landscapes teeming with plant life. However, a crucial component of permaculture often overlooked is the integration of small-scale livestock. This section will explore the benefits of incorporating animals into urban permaculture systems and discuss how to do so effectively and ethically.

First, let's address why animals are included in urban permaculture. The answer lies in the symbiotic relationship between plants and animals. Animals can provide valuable services to your garden, such as pest control, fertilization, and soil aeration. In return, your garden offers these creatures food, shelter, and natural habitat. This mutually beneficial relationship not only enhances the productivity of your garden but also promotes biodiversity and ecological balance.

Now, let's dive into some practical ways to integrate small-scale livestock into your urban permaculture system:

Chickens: These feathered friends are a popular choice for urban permaculture enthusiasts. Chickens can help control pests by eating insects, weeds, and kitchen scraps. They also produce nutrient-rich manure that can be used as fertilizer. Additionally, they provide fresh eggs, a valuable source of protein for your family. When selecting a breed, consider smaller, quieter varieties that are well-suited for urban environments, such as the Bantam or Silkie.

Bees: Urban beekeeping is gaining popularity as more people recognize the importance of pollinators in our ecosystems. Bees not only produce honey but also play a vital role in pollinating plants, ensuring a bountiful harvest. Installing a beehive in your garden can significantly improve the health and productivity of your plants while providing a home for these essential insects.

Rabbits: These small mammals are another excellent addition to your urban permaculture system. Rabbits are efficient at converting plant material into protein and produce high-quality manure that can be used as fertilizer. They can be housed in small hutches and require minimal space, making them ideal for urban settings. Plus, their gentle nature and quiet demeanor make them excellent companions for children and adults alike.

Worms: Vermicomposting, or worm composting, is a fantastic way to recycle kitchen scraps and produce nutrient-rich compost for your garden. Worms break down organic matter, turning it into a valuable soil amendment called "worm castings." This process not only reduces waste but also improves soil structure and fertility. A simple worm bin can be set up in a small space, such as a balcony or garage, making it an ideal option for urban permaculture practitioners.

Fish: Aquaponics is an innovative method of growing plants and fish together in a closed-loop system. The fish waste provides nutrients for the plants, while the plants filter the water, creating a clean environment for the fish. This symbiotic relationship allows for the efficient use of space and resources, making it an attractive option for urban permaculture enthusiasts.

In conclusion, integrating small-scale livestock into your urban permaculture system can provide numerous benefits, from increased garden productivity to enhanced biodiversity. By carefully selecting the right animals for your space and needs, you can create a thriving, sustainable ecosystem that supports both plants and animals. Embrace the transformative power of urban and small-space permaculture by incorporating these living elements into your garden, and watch as your green oasis flourishes.

The Transformative Power of Urban and Small-Space Permaculture

As we reach the end of this enlightening journey through the world of urban and small-space permaculture, it is essential to reflect on the transformative power this practice holds for individuals, communities, and our planet. In a time when urbanization is rapidly expanding and the need for sustainable living practices is more urgent than ever, permaculture offers a beacon of hope and a practical solution to many of the challenges we face.

Throughout this chapter, we have explored various techniques and strategies that enable city dwellers and those with limited space to **embrace permaculture principles**. From vertical gardening and multi-functional elements to urban food forests and community connections, these approaches empower individuals to make the most of their available space while contributing to a more sustainable and resilient future.

Integrating small-scale livestock into urban permaculture systems further demonstrates the adaptability and versatility of this practice. By incorporating animals into our urban ecosystems, we not only benefit from their valuable contributions to soil fertility and pest control, but we also foster a deeper connection to the natural world and the cycles of life that sustains us.

Perhaps the most significant aspect of urban and small-space permaculture is its ability to unite people. As we cultivate local networks and collaborate with our neighbors, we create a sense of community and belonging that is often lacking in modern urban environments. This social aspect of permaculture not only strengthens our resilience in the face of adversity but also contributes to our overall well-being and happiness.

In conclusion, the transformative power of urban and small-space permaculture lies in its capacity to inspire change on both a personal and collective level. By embracing these principles and practices, we can create thriving, sustainable ecosystems that nourish our bodies, minds, and souls, while also contributing to the health and vitality of our planet. As more and more people adopt permaculture practices in their urban and small spaces, we move closer to a world where harmony with nature and sustainable living become the norm rather than the exception. So, let us continue to learn, grow, and share our knowledge with others as we work together to create a brighter, greener future for all.

Chapter Summary

- Urban and small-space permaculture is a powerful approach to sustainable living that can be adapted to fit the unique challenges and opportunities presented by limited spaces.
- Vertical gardening and multi-functional elements are key strategies for maximizing space in urban and small-scale settings, allowing for increased productivity and efficiency in limited areas.
- Urban food forests offer a unique and innovative solution to sustainable food production in cities, transforming underutilized spaces into thriving, edible landscapes.
- Building community connections and fostering local networks are essential components of urban and small-space permaculture, contributing to the resilience and success of permaculture projects.
- Small-scale livestock, such as chickens, bees, rabbits, worms, and fish, can be integrated into urban permaculture systems to provide valuable services, promote biodiversity, and enhance garden productivity.
- Permaculture practices in urban and small spaces can bring people together, creating a sense of community and belonging that is often lacking in modern urban environments.
- The transformative power of urban and small-space permaculture lies in its capacity to inspire personal and

collective change, contributing to a more sustainable and resilient future.

- As more people adopt permaculture practices in their urban and small spaces, we move closer to a world where harmony with nature and sustainable living become the norm rather than the exception.

10

HARVESTING, PRESERVING, AND SHARING THE ABUNDANCE

A magical vegetable waterfall, with a cascade of various vegetables flowing down a mountainside into a lush, green valley, symbolizing the endless abundance of nature's bounty.

Permaculture is a holistic approach to agriculture and sustainable living that seeks to create a harmonious relationship between humans, plants, animals, and the environment. By designing systems that mimic natural ecosystems, permaculture practitioners aim to produce abundant food, fiber, and energy while minimizing waste and nurturing the health of the land. This chapter will delve into the rewarding process of harvesting, preserving, and sharing the fruits of your permaculture labor.

As you immerse yourself in the world of permaculture, you will quickly discover that the bountiful harvests it yields are not only a **source of nourishment** but also **a symbol of the interconnectedness of all living things.** The abundance that springs forth from a well-designed permaculture system is a testament to the power of working with nature rather than against it. By embracing the principles of permaculture, you are not only cultivating a thriving garden but also fostering a deeper connection to the earth and your community.

The process of harvesting, preserving, and sharing the abundance of your permaculture garden is a celebration of the cycles of life and the generosity of nature. As you learn to recognize the optimal time to harvest each plant, you will develop a keen awareness of the rhythms of the natural world. Preserving the fruits of your labor allows you to extend the life of your harvest, ensuring that you can enjoy the flavors and nutrients of your garden throughout the year. Sharing your bounty with friends, family, and neighbors is an opportunity to strengthen social bonds and spread the message of permaculture to a wider audience.

In this chapter, we will explore the art of harvesting, delving into the best practices for timing and techniques to ensure the highest quality and yield. We will then discuss various methods for preserving your harvest, from traditional techniques like canning and drying to innovative approaches such as fermentation and freezing. Finally, we will examine the importance of sharing the wealth of your permaculture garden, highlighting the ways in which this practice can foster community connections and promote a more sustainable future.

As you embark on this journey of harvesting, preserving, and sharing the abundance of your permaculture garden, remember that you are nurturing your well-being and contributing to the planet's health and vitality. Furthermore, by embracing permaculture principles, you are planting the seeds of a more sustainable, just, and abundant future for all.

The Art of Harvesting: Timing and Techniques

In the world of permaculture, the art of harvesting is a crucial skill that every practitioner must master. Knowing when and how to harvest your crops ensures that you reap the maximum benefits from your garden while maintaining the health and vitality of your plants. In this section, we will explore the importance of proper timing and techniques in harvesting, allowing you to fully embrace the abundance of permaculture.

Timing is Everything

One of the most critical aspects of harvesting is knowing when to do it. The ideal time to harvest varies depending on the crop type, the weather, and the specific needs of your garden. Here are some general guidelines to help you determine the best time to harvest your produce:

Observe your plants: Pay close attention to the appearance of your crops. Look for signs of ripeness, such as a change in color, size, or texture. For example, tomatoes are ready to harvest when they have a deep red color and a slightly soft texture.

Taste test: Sampling a small portion of your produce can help you determine if it's ready to harvest. The taste should be flavorful and not bitter or bland.

Monitor the weather: Weather conditions can significantly impact the quality of your harvest. Generally, it's best to harvest in the early morning or late afternoon when temperatures are more relaxed. This helps to preserve the freshness and flavor of your produce.

Consider the plant's lifecycle: Some plants, such as leafy greens, can be harvested multiple times throughout their lifecycle. In these cases, it's essential to harvest in a way that promotes continued growth and production.

Mastering the Techniques

Once you've determined the optimal time to harvest your crops, it's essential to use the proper techniques to ensure the health of your plants and the quality of your produce. Here are some essential harvesting techniques to keep in mind:

Use clean, sharp tools: Dull or dirty tools can damage your plants and introduce harmful pathogens. Therefore, ensure that your harvesting tools, such as knives, scissors, or pruners, are clean and sharp before use.

Be gentle: When harvesting, handle your produce with care to avoid bruising or damaging the fruits and vegetables. This not only preserves the quality of your harvest but also helps to prevent the spread of disease.

Harvest selectively: Instead of harvesting an entire plant at once, consider selectively picking the ripest fruits or leaves. This allows the plant to continue producing and encourages a more extended harvest season.

Leave some behind: In the spirit of permaculture, consider leaving a small portion of your harvest behind for wildlife, insects, and natural decomposition. This helps to support the ecosystem and contributes to the overall health of your garden.

By mastering the art of harvesting, you can ensure that you are making the most of the abundance that your permaculture garden has to offer. By paying close attention to timing and using proper techniques, you can enjoy a bountiful and sustainable harvest that nourishes both your body and the environment.

Preserving Your Harvest: Methods for Longevity and Quality

As the fruits of your permaculture labor begin to ripen and fill your garden with an array of colors, flavors, and textures, it's time to consider the best ways to preserve your harvest. By employing various preservation techniques, you can extend the life of your produce, reduce food waste, and enjoy the fruits of your labor throughout the year. This section will explore several methods for preserving your harvest, ensuring both longevity and quality.

Drying: Drying is one of the oldest and most straightforward methods of preserving food. By removing the moisture from your produce, you inhibit the growth of bacteria and mold, thus prolonging its shelf life. There are several ways to dry your harvest, including air drying, oven drying, and using a food dehydrator. Herbs, fruits, and vegetables can all be dried and stored in airtight containers for future use. Additionally, dried foods are lightweight and easy to transport, making them perfect for camping trips or as emergency provisions.

Canning: Canning is another popular method for preserving your harvest. By sealing your produce in airtight containers and subjecting them to high temperatures, you can effectively kill any bacteria or microorganisms that may cause spoilage. There are two primary methods of canning: water bath canning and pressure canning. Water bath canning is suitable for high-acid foods such as fruits, pickles, and jams, while pressure canning is necessary for low-acid foods like vegetables and meats. Canned goods can last for years when stored in a cool, dark place, providing you with a reliable source of nutritious food.

Freezing: Freezing is a quick and easy way to preserve your harvest, particularly for fruits and vegetables that may not be suitable for drying or canning. By freezing your produce at its peak ripeness, you can lock in its flavor and nutritional content. It's essential to blanch vegetables before freezing to preserve their color, texture, and nutritional value. Fruits can be frozen whole or sliced, with or without added sugar. Remember to use proper freezer-safe containers to prevent freezer burn and ensure the quality of your preserved produce.

Fermenting: Fermentation is a natural process that transforms your harvest into a flavorful, probiotic-rich food. By creating an anaerobic environment, beneficial bacteria can thrive and preserve your produce by converting sugars into lactic acid. Some popular fermented foods include sauerkraut, kimchi, pickles, and yogurt. Fermenting your harvest not only extends its shelf life but also enhances its nutritional content and adds a unique depth of flavor.

Root Cellaring: Root cellaring is a traditional method of storing fruits and vegetables in a cool, dark, and humid environment. This method relies on the earth's natural properties to maintain a stable temperature and humidity level, slowing the ripening process and preserving your harvest for an extended period. Root cellaring is ideal for storing root vegetables, winter squash, apples, and pears. By creating a dedicated space for root cellaring, you can keep your harvest fresh and accessible throughout the winter months.

In conclusion, preserving your harvest is an essential aspect of permaculture, allowing you to enjoy the fruits of your labor year-round and reduce food waste. By employing various preservation techniques, you can ensure the longevity and quality of your produce while adding diversity to your diet. As you continue to explore the world of permaculture, remember that preserving your harvest is just as important as growing it, and together, they contribute to a sustainable and abundant future.

Sharing the Wealth: Building Community through Permaculture

One of the most rewarding aspects of permaculture is the opportunity to share the fruits of your labor with others. By cultivating an environment that fosters collaboration, generosity, and a sense of belonging, you can create a strong, resilient community that thrives on the principles of permaculture. In this section, we will explore the various ways in which you can share the wealth of your permaculture garden and build a vibrant, interconnected community.

Community Gardens and Allotments: A great way to share the abundance of your permaculture garden is by participating in or establishing a community garden or allotment. These shared spaces provide an opportunity for individuals to come together, learn from one another, and collectively work towards a common goal. They also serve as a platform for sharing resources, knowledge, and skills, fostering a sense of camaraderie and mutual support among participants.

Seed and Plant Swaps: Organizing or attending seed and plant swaps is another excellent way to share the wealth of your permaculture garden. These events allow gardeners to exchange seeds, seedlings, and plants, promoting biodiversity and encouraging the cultivation of heirlooms and locally adapted varieties. Seed and plant swaps also provide a forum for sharing knowledge and experiences, helping to strengthen the bonds between community members.

Workshops and Skill Shares: Sharing your knowledge and skills is an invaluable way to contribute to your community and promote the principles of permaculture. By offering workshops, demonstrations, or informal skill shares, you can empower others to take control of their own food production and contribute to developing a more sustainable, self-reliant community. Workshop topics might include composting, natural pest control, water harvesting, or food preservation techniques.

Food Donations and Gleaning Programs: Donating surplus produce to local food banks, shelters, or community kitchens is a wonderful way to share the wealth of your permaculture garden with those in need. Gleaning programs, which involve collecting unharvested produce from farms and gardens to distribute to those experiencing food insecurity, are another effective means of sharing the abundance. Participating in these initiatives can help alleviate hunger and promote food justice within your community.

Community Supported Agriculture (CSA) and Food Co-ops: Joining or establishing a Community Supported Agriculture (CSA) program or food co-op is yet another way to share the wealth of your permaculture garden. These models promote a direct relationship between producers and consumers, fostering a sense of shared responsibility and mutual support. Participating in a CSA or food co-op can help create a more equitable, sustainable, and resilient local food system.

In conclusion, sharing the wealth of your permaculture garden is not only a fulfilling and rewarding endeavor but also an essential component of building a strong, interconnected community. By embracing the principles of collaboration, generosity, and mutual support, you can contribute to cultivating a sustainable future that benefits us all.

Cultivating a Sustainable Future through Permaculture

As we reach the end of our journey through harvesting, preserving, and sharing the abundance of permaculture, it is essential to reflect on the broader implications of these practices for our world. The principles of permaculture not only promote a sustainable and self-sufficient lifestyle but also foster a deep connection with nature, our communities, and ourselves. By embracing the tenets of permaculture, we can cultivate a brighter, more resilient future for generations to come.

Throughout this chapter, we have explored the importance of timing and techniques in harvesting, various methods for preserving the fruits of our labor, and the power of sharing our bounty with others. These practices not only provide us with nourishment and sustenance but also serve as a reminder of our interdependence with the earth and its ecosystems. By working in harmony with nature, we can harness its abundance while minimizing our impact on the environment.

Moreover, sharing our harvest with others fosters a sense of community and connection often lacking in today's fast-paced, consumer-driven society. By engaging in collaborative efforts, such as community gardens or food-sharing initiatives, we can strengthen our bonds with neighbors and friends while promoting a more equitable distribution of resources. This spirit of generosity and reciprocity is at the heart of permaculture and serves as a powerful antidote to the isolation and disconnection that many people experience in the modern world.

In conclusion, the principles and practices of permaculture offer a roadmap for cultivating a sustainable future rooted in abundance, resilience, and community. By embracing these values and integrating them into our daily lives, we can create a world where humans and nature can thrive. As we harvest, preserve, and share the fruits of our labor, we are not only nourishing our bodies but also nourishing our souls and the collective spirit of our communities. Together, we can sow the seeds of change and reap the rewards of a more just, equitable, and sustainable world.

Chapter Summary

- Permaculture is a holistic approach to agriculture and sustainable living that focuses on creating a harmonious relationship between humans, plants, animals, and the environment, resulting in abundant food, fiber, and energy production.
- Proper harvesting techniques and timing are crucial for ensuring the highest quality and yield of produce, as well as maintaining the health and vitality of plants.
- Preserving the harvest through various methods, such as drying, canning, freezing, fermenting, and root cellaring, allows for the enjoyment of garden produce throughout the year and reduces food waste.
- Sharing the abundance of a permaculture garden with friends, family, and neighbors strengthens social bonds and promotes the principles of permaculture to a wider audience.
- Community gardens, seed and plant swaps, workshops, food donations, and gleaning programs are all ways to share the wealth of a permaculture garden and build a strong, interconnected community.
- The principles of permaculture not only promote a sustainable and self-sufficient lifestyle but also foster a deep connection with nature, our communities, and ourselves,

contributing to a brighter, more resilient future for generations to come.

- By embracing the tenets of permaculture and integrating them into our daily lives, we can create a world where both humans and nature can thrive, fostering a sense of community, generosity, and reciprocity.

THE FUTURE OF PERMACULTURE

As we conclude our permaculture journey, it is essential to take a moment to reflect on the knowledge and insights we have gained throughout this exploration. As we have discovered, permaculture is not merely a set of gardening techniques or a trendy buzzword. It is a holistic approach to living in harmony with nature, creating sustainable systems that benefit both people and the planet. By embracing permaculture principles, we can transform our lives and our environment for the better.

Throughout this book, we have delved into the core concepts of permaculture, examining its ethical foundation and the practical applications of its principles. We have explored the importance of observing and understanding natural patterns and the value of designing systems that mimic these patterns to create resilient and productive ecosystems. From soil health and water management to plant diversity and animal integration, we have learned how permaculture practices can be applied to various aspects of our lives, whether we live on a small urban plot or a sprawling rural farm.

As we have seen, permaculture is not a one-size-fits-all solution but a **flexible framework** that can be adapted to suit individual needs and local conditions. By adopting a permaculture mindset, we can become more mindful of our actions and their consequences, and make informed decisions that promote sustainability and regeneration. This journey has shown us that permaculture is not just about growing food but about cultivating a deeper connection with the natural world and fostering a sense of responsibility for its well-being.

In the following sections, we will explore the broader implications of permaculture for our society and planet. We will discuss how permaculture is shaping the future of agriculture, offering innovative solutions to global environmental challenges, and inspiring a new generation of eco-conscious citizens. We will also examine the role of technology and innovation in advancing permaculture practices and consider how education and community involvement can help to spread the permaculture ethos.

As we conclude our permaculture journey, let us remember that this is not the end but the beginning of a lifelong adventure. By embracing the principles of permaculture, we can continue to learn, grow, and evolve, both as individuals and as a collective, working together to create a more sustainable and harmonious future for all.

The Growing Impact of Permaculture on Modern Agriculture

As we have journeyed through the world of practical permaculture, it has become increasingly evident that this holistic approach to agriculture is not only beneficial for our gardens and homes but also holds the potential to revolutionize modern agriculture. The growing impact of permaculture on today's farming practices is a testament to its effectiveness and sustainability as more and more farmers and agriculturalists embrace its principles.

One of the most significant ways permaculture influences modern agriculture is by promoting **biodiversity** and **ecological balance.** Traditional farming methods often rely on monoculture, which involves the cultivation of a single crop in a given area. While this approach may yield high production levels in the short term, it can lead to soil depletion, increased vulnerability to pests, and a decline in overall ecosystem health. In contrast, permaculture encourages the cultivation of diverse plant species, which not only fosters a more resilient and productive system but also supports the health of the surrounding environment.

Another notable impact of permaculture on modern agriculture is its focus on **regenerative practices.** Conventional farming methods often deplete the soil and contribute to environmental degradation through synthetic fertilizers, pesticides, and intensive tillage. Permaculture, on the other hand, emphasizes the importance of nurturing the soil and preserving its natural fertility. By employing techniques such as composting, mulching, and cover cropping, permaculture practitioners work to replenish the soil's nutrients and promote its long-term health. This shift towards regenerative agriculture not only enhances the land's productivity but also helps mitigate the adverse effects of climate change by sequestering carbon in the soil.

Furthermore, permaculture's emphasis on local, small-scale production is gradually transforming the way we think about agriculture and food systems. In an era of industrial agriculture and globalized food chains, the importance of local, sustainable food production cannot be overstated. Permaculture encourages the development of community-supported agriculture (CSA) initiatives, farmers' markets, and urban gardens, which not only reduce the environmental footprint of our food systems but also foster stronger connections between producers and consumers.

Finally, the growing impact of permaculture on modern agriculture can be seen in the increasing number of educational programs and resources dedicated to its principles. From workshops and online courses to books and documentaries, a wealth of information is available for those interested in learning more about permaculture and its applications. This widespread dissemination of knowledge is crucial for the continued growth and development of permaculture as a viable alternative to conventional agriculture.

In conclusion, the growing impact of permaculture on modern agriculture is a testament to its effectiveness, sustainability, and potential to address some of the most pressing challenges facing our global food systems. By embracing permaculture principles, we can work towards a more resilient, ecologically balanced, and sustainable future for agriculture and our planet.

Permaculture as a Solution to Global Environmental Challenges

As we stand at the precipice of unprecedented environmental challenges, it becomes increasingly evident that our current agricultural practices and consumption patterns are unsustainable. Climate change, deforestation, soil degradation, and water scarcity are just a few of the pressing issues that threaten the future of our planet. In this context, permaculture emerges as a beacon of hope, offering a holistic and regenerative approach to agriculture that can help mitigate these challenges and pave the way for a more sustainable future.

One of the core tenets of permaculture is its emphasis on **working with**, rather than against, **nature.** By mimicking natural ecosystems and promoting biodiversity, permaculture practices can help restore degraded lands and foster resilience in the face of climate change. For instance, agroforestry techniques, which involve integrating trees and shrubs into agricultural systems, can help sequester carbon, reduce soil erosion, and provide a habitat for beneficial insects and wildlife. Similarly, the practice of intercropping, or growing multiple crops together, can increase the overall productivity of a plot while also improving soil health and reducing the need for chemical inputs.

Another critical aspect of permaculture is its focus on **closed-loop systems** that minimize waste and make the most of available resources. This approach reduces the environmental footprint of agricultural activities and helps address issues of food security and resource scarcity. For example, composting and vermiculture (worm farming) can transform organic waste into valuable fertilizer, reducing the need for synthetic inputs and diverting waste from landfills. Moreover, permaculture encourages the use of water-saving techniques, such as rainwater harvesting, swales, and mulching, which can help conserve water resources and adapt to changing precipitation patterns.

In addition to its direct environmental benefits, permaculture also has the potential to **foster social change** by promoting a more equitable and community-oriented approach to food production. By emphasizing local, small-scale, and diverse agricultural systems, permaculture can help counteract the negative impacts of industrial agriculture, such as the loss of traditional knowledge, the concentration of land ownership, and the marginalization of small-scale farmers. Furthermore, by encouraging community involvement and education, permaculture can empower individuals to take control of their food systems and become active agents of change.

In conclusion, permaculture offers a powerful and practical solution to many of the global environmental challenges we face today. By embracing its principles and practices, we can create more resilient and sustainable agricultural systems and foster a deeper connection with the natural world and our communities. As we look towards the future, it is clear that permaculture has a vital role to play in cultivating a more just, sustainable, and thriving planet for generations to come.

The Role of Technology and Innovation in Advancing Permaculture Practices

As we venture into the future, technology and innovation will play a crucial role in advancing permaculture practices. By harnessing the power of cutting-edge tools and techniques, we can optimize our efforts to create sustainable, resilient, and productive ecosystems that benefit both people and the planet.

One of the most promising areas of technological advancement is the field of **precision agriculture.** This approach uses data-driven insights and advanced monitoring systems to make more informed decisions about managing land, water, and other resources. By integrating these tools into permaculture practices, we can enhance our ability to observe, analyze, and respond to the unique needs of each site, ultimately leading to more efficient and effective systems.

For example, remote sensing technology, such as drones and satellite imagery, can provide valuable information about soil health, plant growth, and water usage. By analyzing this data, permaculture practitioners can identify patterns and trends that inform their design decisions, leading to more targeted interventions that maximize productivity while minimizing waste and environmental impact.

Another area of innovation is the development of new materials and techniques for **sustainable construction** and **infrastructure.** From green roofs and living walls to permeable paving and natural water filtration systems, these technologies offer exciting possibilities for integrating permaculture principles into the built environment. By embracing these innovations, we can create urban landscapes that support human well-being and contribute to the health and vitality of the ecosystems in which they are situated.

Furthermore, the rise of **digital communication** and **social media platforms** has created unprecedented opportunities for sharing knowledge, ideas, and best practices within the global permaculture community. By leveraging these tools, we can foster a spirit of collaboration and mutual learning that accelerates the spread of permaculture principles and techniques around the world.

Finally, it is important to recognize that technology and innovation are not just about the latest gadgets and gizmos. They also encompass the creative thinking and problem-solving skills that are at the heart of permaculture design. By cultivating a mindset of curiosity, experimentation, and adaptability, we can continually refine and improve our practices, ensuring that permaculture remains a dynamic and evolving discipline that is well-equipped to meet the challenges of the future.

In conclusion, the integration of technology and innovation into permaculture practices holds immense potential for enhancing the sustainability, resilience, and productivity of our ecosystems. By embracing these advancements and fostering a culture of continuous learning and improvement, we can ensure that permaculture remains at the forefront of efforts to create a more just, equitable, and ecologically sound world for future generations.

Fostering a Permaculture Mindset: Education and Community Involvement

As we have journeyed through the world of practical permaculture, it has become evident that the key to unlocking its full potential lies in fostering a permaculture mindset. This mindset, rooted in the principles of sustainability, resilience, and harmony with nature, can only be cultivated through education and community involvement. By spreading awareness and knowledge about permaculture practices, we can empower individuals and communities to take charge of their own food systems and contribute to a more sustainable future.

Education is the cornerstone of fostering a permaculture mindset. We can lay the foundation for a new generation of eco-conscious individuals by incorporating permaculture principles into school curriculums. Teaching children about the importance of soil health, water conservation, and biodiversity will instill a deep appreciation for the natural world and humans' role in preserving it. Furthermore, offering workshops and courses for adults can also broaden the reach of permaculture, equipping people with the skills and knowledge necessary to implement these practices in their own lives.

Community involvement is another essential aspect of nurturing a permaculture mindset. By creating spaces where people can come together to learn, share, and collaborate, we can foster a sense of unity and collective responsibility for the well-being of our planet. Community gardens, for example, serve as a tangible representation of permaculture principles in action, providing a platform for individuals to work together in cultivating a sustainable food system. These gardens not only produce fresh, local produce but also serve as a hub for learning and social interaction, strengthening the bonds between community members.

In addition to community gardens, local workshops, seminars, and conferences can also serve as opportunities for individuals to connect with like-minded people and learn from experts in the field. Community members can gain valuable insights and inspiration by engaging in these events to further their permaculture journey. Moreover, forming local permaculture groups or clubs can provide ongoing support and resources for those looking to deepen their understanding and practice of permaculture principles.

Ultimately, fostering a permaculture mindset is about more than just adopting specific techniques or practices. It is about embracing a holistic approach to living, one that recognizes the interconnectedness of all living systems and seeks to create harmony between human needs and the natural world. By prioritizing education and community involvement, we can cultivate a permaculture mindset that will not only benefit our immediate surroundings but also contribute to the global movement toward a more sustainable and resilient future.

Cultivating a Sustainable Future: The Lasting Legacy of Permaculture

As we stand at the precipice of a new era, it is crucial to recognize the importance of sustainable practices in shaping the future of our planet. With its emphasis on working in harmony with nature, permaculture offers a viable and promising solution to many of the challenges we face today. By embracing permaculture principles, we can cultivate a sustainable future that benefits both the environment and humanity.

The lasting legacy of permaculture lies in its ability to transform our relationship with the natural world. By adopting a holistic approach to agriculture, we can create self-sustaining ecosystems that provide for our needs while preserving the integrity of the environment. This shift in mindset, **from exploitation to stewardship**, can revolutionize how we interact with the earth and its resources.

Permaculture's impact extends beyond the realm of agriculture, as its principles can be applied to various aspects of our daily lives. From energy-efficient housing to waste reduction and water conservation, adopting permaculture practices can significantly reduce our ecological footprint and contribute to a more sustainable future.

Integrating technology and innovation in permaculture practices further enhances its potential to bring about lasting change. By harnessing the power of cutting-edge tools and techniques, we can optimize the efficiency of permaculture systems and maximize their positive impact on the environment. This marriage of traditional wisdom and modern innovation is a testament to the adaptability and resilience of permaculture as a solution for the challenges of the *21st century*.

Education and community involvement play a crucial role in fostering a permaculture mindset. By raising awareness about the principles of permaculture and their practical applications, we can inspire individuals and communities to take action toward a more sustainable future. This grassroots movement can create a ripple effect, spreading the message of permaculture far and wide and ultimately transforming how we live on this planet.

In conclusion, the future of permaculture is one of hope and promise. As we continue to face unprecedented environmental challenges, adopting sustainable practices like permaculture becomes increasingly vital. By embracing the principles of permaculture, we can cultivate a sustainable future that leaves a lasting legacy for generations to come. Together, we can work towards a world where humans and nature coexist in harmony, ensuring the health and well-being of our planet and all its inhabitants.

AGROFORESTRY AND SILVOPASTURE

A BEGINNER'S GUIDE TO UNITING LIVESTOCK, TREES, AND COMMUNITIES TO COMBAT CLIMATE CHANGE AND BOOST BIODIVERSITY

INTRODUCTION

In today's rapidly changing world, the need for sustainable and efficient agricultural practices has never been more critical. With a growing global population and increasing pressure on our planet's finite resources, it is essential to explore innovative solutions that can help meet food, fuel, and fiber demands while preserving the environment. Agroforestry and silvopasture are two such solutions that have the potential to revolutionize the way we approach agriculture and land management. This chapter will explore the fascinating world of agroforestry and silvopasture, exploring their origins, principles, and benefits.

Agroforestry, at its core, is the integration of trees and shrubs into agricultural systems. This practice combines the best of both worlds: the productivity and efficiency of agriculture with the ecological benefits of forests. By incorporating trees into farming systems, agroforestry can provide many benefits, including improved soil fertility, increased biodiversity, enhanced water quality, and reduced erosion. Moreover, agroforestry systems can also provide valuable products such as timber, fruits, nuts, and medicinal plants, diversifying farmers' income sources and contributing to food security.

Silvopasture, a subset of agroforestry, is the practice of integrating trees, livestock, and forage production on the same land. This system offers a unique opportunity to maximize the use of available resources, as trees provide shade and shelter for livestock, while their roots help to maintain soil structure and prevent erosion. In turn, livestock can help manage vegetation, fertilize the soil, and control pests. The combination of these elements creates a dynamic, resilient, and productive ecosystem that can support both agricultural and environmental goals.

The concept of agroforestry and silvopasture is not new; in fact, it has been practiced for centuries by indigenous communities around the world. However, the modern agricultural revolution, with its focus on monoculture and intensive farming practices, has led to a decline in the use of these traditional systems. As we face the challenges of climate change, resource scarcity, and environmental degradation, there is a renewed interest in agroforestry and silvopasture as sustainable alternatives to conventional agriculture.

In the following sections, we will explore the historical context of agroforestry and silvopasture, delving into their origins and evolution over time. We will then examine the fundamental principles and components of agroforestry systems, highlighting the various techniques and strategies that can be employed to create diverse, resilient, and productive landscapes. Next, we will focus on silvopasture, discussing its benefits and techniques in greater detail. Finally, we will present case studies of successful agroforestry and silvopasture implementations from around the world, showcasing the potential of these systems to transform agriculture and contribute to a more sustainable future.

Join us on this exciting journey as we uncover the potential of agroforestry and silvopasture and discover how these innovative practices can help to shape the future of sustainable agriculture.

A Brief Historical Context of Agroforestry and Silvopasture

The roots of agroforestry and silvopasture can be traced back thousands of years, with evidence of these practices found in various ancient civilizations across the globe. To fully appreciate the significance of these sustainable agricultural systems, it is essential to delve into their historical context and understand how they have evolved.

Agroforestry, the intentional integration of trees and shrubs into crop and animal farming systems, has been practiced for millennia. Ancient civilizations, such as the Maya in Central America, the indigenous people of the Amazon Basin, and various African tribes, recognized the benefits of combining trees with agriculture. These early societies understood that trees played a crucial role in maintaining soil fertility, preventing erosion, and providing shade and habitat for wildlife. They also valued the diverse array of products that trees could provide, such as fruits, nuts, timber, and medicine.

Similarly, silvopasture, which involves the integration of trees, livestock, and forage plants, has a long and rich history. In Europe, silvopasture can be traced back to the Roman Empire. During this period, farmers grazed their livestock in wooded areas to take advantage of the natural forage and shelter provided by the trees. This practice continued throughout the Middle Ages and is still common in many parts of Europe today.

In Asia, traditional agroforestry systems, such as the home gardens of Indonesia and the tree-based agriculture of the Philippines, have been practiced for centuries. These systems not only provided food and income for local communities but also played a vital role in preserving biodiversity and maintaining ecological balance.

Despite their ancient origins, agroforestry and silvopasture experienced a decline in popularity during the *20th century,* as modern agricultural practices prioritized monoculture and the use of synthetic fertilizers and pesticides. However, in recent decades, there has been a resurgence of interest in these sustainable agricultural systems, driven by a growing awareness of their environmental, social, and economic benefits.

Today, agroforestry and silvopasture are recognized as valuable tools in the quest for sustainable agriculture. Numerous organizations, researchers, and farmers are working to promote adoption and further refine their techniques. By understanding the historical context of these practices, we can appreciate their enduring relevance and potential to shape the future of agriculture in a more sustainable and resilient direction.

Key Principles and Components of Agroforestry Systems

Agroforestry, a sustainable land management practice, combines agriculture and forestry to create integrated, diverse, productive, and resilient ecosystems. This innovative approach to land use offers numerous benefits, including increased biodiversity, improved soil health, and enhanced economic opportunities for farmers. This section will delve into the fundamental principles and components of successful agroforestry systems.

Diversity: Agroforestry systems thrive on diversity, both in terms of plant species and agricultural practices. By incorporating a variety of trees, shrubs, and crops, these systems create a more resilient ecosystem that can better withstand pests, diseases, and climate fluctuations. This diversity also provides a range of products and services, such as food, timber, and habitat for wildlife, which can contribute to the overall sustainability and profitability of the system.

Integration: The integration of trees, crops, and livestock is a fundamental aspect of agroforestry systems. This integration allows for the efficient use of resources, such as sunlight, water, and nutrients, as well as the recycling of organic matter. Trees can provide shade and shelter for crops and livestock, while their roots help to improve soil structure and prevent erosion. In turn, crops and livestock can contribute to the system's overall productivity by providing food, fiber, and other valuable products.

Synergy: Agroforestry systems are designed to maximize the positive interactions between trees, crops, and livestock. These synergies can increase productivity, improve resource use efficiency, and enhance ecosystem services. For example, trees can help to improve soil fertility by fixing nitrogen, while livestock can help to control weeds and pests through grazing. By working together, the various components of an agroforestry system can create a more sustainable and productive landscape.

Adaptability: Agroforestry systems are inherently adaptable, as they can be tailored to suit various environmental conditions, cultural practices, and economic objectives. This adaptability allows farmers to experiment with different tree species, crop rotations, and livestock management strategies to find the best fit for their unique circumstances. By being flexible and responsive to change, agroforestry systems can help to build resilience in the face of climate change and other environmental challenges.

Sustainability: The ultimate goal of agroforestry is to create sustainable agricultural systems that provide long-term benefits for both people and the environment. By promoting biodiversity, conserving natural resources, and enhancing ecosystem services, agroforestry systems can contribute to the overall health and resilience of our landscapes. Additionally, by diversifying farm income and providing new market opportunities, agroforestry can help to support the economic viability of rural communities.

In conclusion, the fundamental principles and components of agroforestry systems – diversity, integration, synergy, adaptability, and sustainability – work together to create a holistic approach to land management that benefits both people and the environment. By understanding and applying these principles, farmers, landowners, and policymakers can help to promote the widespread adoption of agroforestry and contribute to a more sustainable future for agriculture.

Understanding Silvopasture: Benefits and Techniques

Silvopasture, a vital component of agroforestry, is a sustainable land management system that combines the cultivation of trees, forage plants, and livestock in a mutually beneficial environment. This innovative approach to agriculture offers numerous advantages for both the ecosystem and the farmers who practice it. In this section, we will delve into the benefits of silvopasture and explore the techniques used to create a successful silvopastoral system.

Environmental Benefits: Silvopasture systems contribute to the preservation of biodiversity, soil health, and water quality. Integrating trees and forage plants helps prevent soil erosion, improve nutrient cycling, and reduce water runoff. Additionally, trees provide habitats for various species of birds, insects, and other wildlife, promoting a more diverse and balanced ecosystem.

Climate Change Mitigation: Trees in silvopasture systems act as carbon sinks, absorbing and storing atmospheric carbon dioxide, which helps to mitigate the effects of climate change. Moreover, the shade provided by trees reduces the need for livestock to expend energy on regulating body temperature, resulting in lower animal methane emissions.

Economic Benefits: Silvopasture systems can provide multiple income streams for farmers, as they can profit from selling timber, non-timber forest products, and livestock. Diversifying income sources reduces the financial risks associated with traditional monoculture farming and can increase overall profitability.

Improved Animal Welfare: Trees in silvopasture systems offer shelter and shade for livestock, protecting them from harsh weather conditions and reducing heat stress. This improved environment results in healthier animals, which can lead to higher productivity and better-quality meat, milk, and other animal products.

Techniques for Implementing Silvopasture

Tree Selection: Choosing the right tree species is crucial for a successful silvopasture system. Some factors to consider are the tree's growth rate, shade tolerance, and compatibility with the forage plants and livestock in the system. Native tree species are often preferred, as they are well-adapted to local conditions and can provide additional benefits for wildlife.

Planting and Spacing: The arrangement and density of trees in a silvopasture system will depend on the farmer's specific goals and the livestock's needs. Trees can be planted in rows, clusters, or scattered throughout the pasture. Proper spacing is essential to ensure adequate sunlight for forage plants and facilitate livestock movement.

Forage Management: Silvopasture systems require careful management of forage plants to maintain productivity and prevent overgrazing. Rotational grazing, in which livestock are moved between different paddocks or sections of the pasture, can help to promote even grazing and allow forage plants to recover.

Livestock Management: The choice of livestock species and the stocking density will depend on the specific objectives of the silvopasture system and the available resources. It is essential to monitor the health and well-being of the animals and to adjust management practices as needed to ensure their welfare and productivity.

In conclusion, silvopasture offers a promising solution for sustainable agriculture, providing environmental, economic, and animal welfare benefits. By understanding and implementing the appropriate techniques, farmers can create thriving silvopastoral systems that contribute to a more sustainable and resilient agricultural landscape.

Case Studies: Successful Agroforestry and Silvopasture Implementations

In this section, we will delve into real-life examples of agroforestry and silvopasture systems that have been successfully implemented around the world. These case studies showcase the potential of these sustainable agricultural practices to improve the livelihoods of farmers, enhance biodiversity, and contribute to climate change mitigation. Furthermore, by examining these success stories, we can gain valuable insights into the practical application of agroforestry and silvopasture principles and techniques.

Case Study 1: The Vi Agroforestry Project in East Africa

The Vi Agroforestry Project, initiated in *1983*, has been working with smallholder farmers in Kenya, Tanzania, Uganda, and Rwanda to promote agroforestry practices. The project has reached over 1.8 million people and helped plant more than 120 million trees. By integrating trees into their farms, the farmers have diversified their income sources, improved soil fertility, and enhanced food security. The project has also contributed to climate change mitigation by sequestering an estimated 2.5 million tons of carbon dioxide.

Case Study 2: Silvopasture in the Dehesa System of Spain and Portugal

The Dehesa system, a traditional land-use practice in Spain and Portugal, is an excellent example of silvopasture. In this system, livestock grazes on grasslands interspersed with oak trees, primarily holm oaks and cork oaks. The trees provide shade and fodder for the animals, while their acorns are a valuable food source for the famous Iberian pigs. The Dehesa system supports a diverse range of flora and fauna, and it has been recognized for its role in preserving the Mediterranean landscape and its cultural heritage.

Case Study 3: The Quesungual Slash and Mulch Agroforestry System in Honduras

The Quesungual Slash and Mulch Agroforestry System (QSMAS) was developed by smallholder farmers in western Honduras to respond to land degradation and food insecurity challenges. This innovative system involves the selective pruning of trees and shrubs, which are then used as mulch to improve soil fertility and moisture retention. The QSMAS has successfully increased crop yields, reduced the need for chemical inputs, and enhanced biodiversity. It has also been recognized for its potential to contribute to climate change adaptation and mitigation.

Case Study 4: Alley Cropping in the United States

Alley cropping is an agroforestry practice that involves planting rows of trees or shrubs alongside annual or perennial crops. In the United States, alley cropping has been successfully implemented in various regions, including the Midwest and the Southeast. For example, in Missouri, farmers have planted rows of black walnut trees with soybeans, corn, and wheat. This system provides multiple benefits, such as improved soil health, increased biodiversity, and additional timber and nut production income.

These case studies demonstrate the versatility and adaptability of agroforestry and silvopasture systems across different regions and ecosystems. By learning from these successful implementations, we can continue to refine and expand the use of these sustainable agricultural practices, paving the way for a more resilient and environmentally friendly future for agriculture.

The Future of Sustainable Agriculture through Agroforestry and Silvopasture

As we conclude our journey through the introduction to agroforestry and silvopasture, it is essential to take a step back and appreciate the immense potential these practices hold for the future of sustainable agriculture. In a world grappling with climate change, environmental degradation, and food security challenges, agroforestry and silvopasture offer innovative solutions that can **revolutionize** the way we approach agriculture and land management.

Throughout this introduction, we have explored the historical context of agroforestry and silvopasture, delved into their fundamental principles and components, and examined the numerous benefits and techniques associated with these practices. We have also looked at real-life case studies that demonstrate the successful implementation of agroforestry and silvopasture systems around the world.

As we look to the future, it is clear that agroforestry and silvopasture have the potential to play a significant role in addressing some of the most pressing challenges of our time. By integrating trees and livestock into agricultural systems, we can enhance biodiversity, improve soil health, increase carbon sequestration, and promote more efficient use of resources. These practices also offer economic benefits to farmers by diversifying income streams, reducing risk, and increasing overall productivity.

However, realizing the full potential of agroforestry and silvopasture will require concerted efforts from various stakeholders, including farmers, researchers, policymakers, and consumers. Investing in research and development is crucial to advance our understanding of these practices and develop innovative techniques that maximize their benefits. Policymakers must create enabling environments that support the adoption of agroforestry and silvopasture through incentives, education, and extension services. Consumers also have a role to play in driving demand for sustainably produced products, thereby encouraging more farmers to adopt these practices.

In conclusion, agroforestry and silvopasture represent a promising pathway toward a more sustainable and resilient agricultural future. By harnessing the power of nature and working in harmony with the environment, we can create food systems that nourish both people and the planet. So, as we move forward, let us embrace the principles of agroforestry and silvopasture and work together to build a brighter, greener future for generations to come.

1

THE HISTORY AND EVOLUTION OF AGROFORESTRY

A futuristic agroforestry landscape, merging the wisdom of traditional practices with cutting-edge technology, to demonstrate the potential of sustainable agriculture in the future.

Agroforestry, the practice of integrating trees and shrubs into agricultural systems, has been a cornerstone of human civilization for thousands of years. As societies evolved and agricultural practices advanced, the benefits of combining trees with traditional farming methods became increasingly apparent. This chapter delves into the rich history of agroforestry, tracing its roots from ancient civilizations to modern innovations and challenges.

The concept of **agroforestry is not new**; in fact, it has been practiced by various cultures across the globe since the dawn of agriculture. Integrating trees and crops or livestock provides numerous benefits, such as improved soil fertility, enhanced biodiversity, and increased resilience against climate change. These advantages have been recognized and harnessed by our ancestors, who developed sophisticated agroforestry systems that laid the foundation for modern practices.

In this chapter, we will explore the historical roots of agroforestry, beginning with ancient civilizations that employed these techniques to sustain their communities and foster environmental stewardship. We will then examine the development of silvopasture systems, which combine forestry and grazing in a mutually beneficial relationship. As we move through the timeline, we will discuss the innovations and challenges that have shaped modern agroforestry and how these practices play a crucial role in sustainable agriculture and climate change mitigation.

By understanding the history and evolution of agroforestry, we can appreciate the wisdom of our ancestors and the importance of continuing to develop and refine these practices for the benefit of future generations. Furthermore, as we face unprecedented environmental challenges, the lessons from the past can serve as a guiding light, illuminating the path toward a sustainable and resilient future for agriculture and our planet.

Ancient Agroforestry Practices and Civilizations

The roots of agroforestry can be traced back to the dawn of human civilization when our ancestors began cultivating crops and domesticating animals. As societies developed, so did their understanding of the complex relationships between plants, animals, and the environment. This section will explore the ancient agroforestry practices and civilizations that laid the foundation for modern agroforestry and silvopasture systems.

One of the earliest examples of agroforestry is found in the indigenous cultures of the Amazon Basin. The Amazonian people practiced a form of agroforestry known as *"forest gardening,"* which involved cultivating multiple layers of vegetation, including trees, shrubs, and herbaceous plants. This diverse and sustainable system provided food, medicine, and other resources for the community while maintaining the integrity of the surrounding ecosystem.

In Africa, agroforestry has been practiced for thousands of years, with evidence of tree cultivation dating back to *6000 BCE*. Ancient African societies recognized the benefits of integrating trees into their agricultural systems, such as improving soil fertility, providing shade for crops, and serving as a source of food and medicine. The practice of planting *Faidherbia albida*, a nitrogen-fixing tree species, alongside cereal crops is a prime example of traditional African agroforestry.

The ancient Mayan civilization in Central America also utilized agroforestry techniques in their agricultural systems. The Mayans cultivated various crops, including maize, beans, squash, and cacao, often intercropping them with fruit and nut trees. This approach maximized the use of available land and promoted biodiversity and soil health.

In Asia, the traditional rice paddy systems of China, Indonesia, and the Philippines incorporated trees and fish into their agricultural landscapes. These integrated systems provided multiple benefits, such as erosion control, nutrient cycling, and increased food production. Similarly, the ancient terraced agroforestry systems of the Andean region in South America combined trees, crops, and livestock to create a sustainable and productive landscape.

As these ancient civilizations demonstrate, agroforestry has been integral to human societies for millennia. The knowledge and techniques developed by our ancestors have been passed down through generations, shaping the foundation of modern agroforestry and silvopasture practices. By understanding and appreciating the historical roots of agroforestry, we can better comprehend its significance in contemporary agriculture and the potential it holds for addressing the challenges of climate change and sustainable food production.

The Development of Silvopasture Systems

Silvopasture, a practice that combines forestry and the grazing of domesticated animals, has a rich and diverse history that has evolved over time. This section will delve into the development of silvopasture systems, exploring their origins, their various forms, and how they have been adapted to suit different environments and cultures.

The roots of silvopasture can be traced back to ancient civilizations, where the integration of trees and livestock was a common practice. In ancient Rome, for example, farmers recognized the benefits of combining trees with pasturelands. The trees provided shade and shelter for the animals while improving soil fertility. Similarly, in pre-Columbian America, indigenous peoples practiced a form of silvopasture known as the *"chinampa"* system, which involved the cultivation of crops on raised beds within shallow lake areas, with trees planted alongside to provide fodder for animals.

As human populations grew and agricultural practices expanded, silvopasture systems continued to evolve. In medieval Europe, the practice of *"wood-pasture"* emerged, where livestock grazed in forests and woodlands, benefiting from the diverse forage and shelter provided by the trees. This system was particularly prevalent in England, where it played a crucial role in shaping the landscape and supporting rural communities.

The geographical and climatic conditions of different regions also influenced the development of silvopasture systems. In the arid and semi-arid regions of Africa and Asia, for example, pastoralists developed systems that integrated trees and shrubs with grazing lands to provide shade, shelter, and fodder for their animals. These systems, known as *"parklands,"* allowed communities to maintain their livestock even in harsh environments with limited resources.

In more recent times, the modernization of agriculture and the rise of industrial farming practices led to a decline in traditional silvopasture systems. However, in the latter part of the *20th century,* there was a resurgence of interest in these systems to promote sustainable agriculture and address environmental challenges. This renewed interest has led to the development of innovative silvopasture techniques, such as planting fast-growing tree species for fodder production, using rotational grazing to improve soil health, and integrating agroforestry practices with modern livestock management.

Today, silvopasture systems are recognized for their potential to **enhance agricultural productivity**, **improve animal welfare**, and contribute to **climate change mitigation.** By sequestering carbon in the soil and trees, these systems can help to offset greenhouse gas emissions from livestock production. Furthermore, silvopasture can play a critical role in promoting biodiversity and preserving ecosystems, as the integration of trees and shrubs with grazing lands creates habitats for a wide range of flora and fauna.

In conclusion, the development of silvopasture systems has been shaped by a complex interplay of historical, cultural, and environmental factors. As we look to the future, it is clear that these systems will continue to evolve and adapt, playing a vital role in pursuing sustainable agriculture and mitigating climate change.

Modern Agroforestry: Innovations and Challenges

In recent years, agroforestry has experienced a resurgence as a viable and sustainable land management practice. This renewed interest is driven by the need for innovative solutions to address pressing global challenges, such as food security, climate change, and environmental degradation. In this section, we will explore the modern innovations in agroforestry and the challenges faced in implementing these practices.

One of the most significant innovations in modern agroforestry is the development of new, highly productive **tree species and cultivars.** These improved varieties have been bred for specific traits, such as rapid growth, high biomass production, and resistance to pests and diseases. This has allowed for the integration of trees into agricultural systems with minimal impact on crop yields while providing additional benefits such as timber, fuelwood, and non-timber forest products.

Another innovation in agroforestry is using advanced spatial planning and design tools. **Geographic Information Systems (GIS)** and **remote sensing** technologies have enabled researchers and practitioners to map and analyze complex agroforestry systems. This allows a better understanding of the interactions between trees, crops, and livestock. This has led to the development of more efficient and sustainable land-use strategies, such as optimizing tree placement to maximize benefits and minimize competition with crops.

In addition to these technological advancements, modern agroforestry has also benefited from a growing body of scientific research. This research has helped to quantify the ecological, economic, and social benefits of agroforestry practices, providing evidence-based support for their adoption and implementation. For example, studies have shown that agroforestry systems can **increase biodiversity, improve soil health**, and **enhance ecosystem services** such as carbon sequestration and water regulation.

Despite these innovations, several challenges still need to be addressed to fully realize the potential of agroforestry. One of the primary challenges is the lack of awareness and understanding of agroforestry practices among farmers, policymakers, and the general public. This can lead to skepticism and resistance to adopting new land management practices, even when they have been proven to be beneficial.

Another challenge is the need for supportive policies and incentives to encourage the adoption of agroforestry practices. In many countries, agricultural policies and subsidies are still heavily focused on monoculture systems, which can discourage farmers from diversifying their land use. To overcome this barrier, governments and international organizations must develop and implement policies that recognize and reward the benefits of agroforestry systems.

Lastly, there is a need for continued research and innovation to address the specific challenges faced by different agroforestry systems in various regions and contexts. This includes developing new tree species and cultivars that are adapted to changing climate conditions, as well as refining management practices to optimize productivity and sustainability.

In conclusion, modern agroforestry has come a long way from its ancient roots, with significant innovations in tree breeding, spatial planning, and scientific research. However, there are still challenges to be addressed in order to fully harness the potential of agroforestry as a sustainable land management practice. By overcoming these challenges and continuing to innovate, agroforestry can play a crucial role in addressing global food security, climate change, and environmental degradation.

The Role of Agroforestry in Sustainable Agriculture and Climate Change Mitigation

As the world grapples with the pressing issues of climate change and food security, agroforestry has emerged as a vital solution that addresses both concerns. By integrating trees into agricultural landscapes, agroforestry promotes sustainable agriculture and plays a significant role in mitigating the effects of climate change. This section delves into the various ways in which agroforestry contributes to these global challenges.

First and foremost, agroforestry enhances the productivity and resilience of agricultural systems. The presence of trees in farmlands provides numerous benefits, such as improving **soil fertility, reducing soil erosion**, and **regulating microclimates.** These factors contribute to increased crop yields and a more stable food supply, which is crucial in the face of a growing global population. Furthermore, the diversification of income sources through agroforestry, such as timber and non-timber forest products, strengthens the economic resilience of farmers and rural communities.

In terms of climate change mitigation, agroforestry systems have a remarkable capacity to sequester carbon. Trees absorb and store atmospheric carbon dioxide, effectively reducing greenhouse gas emissions. Studies have shown that agroforestry systems can sequester up to four times more carbon than conventional agricultural systems. This carbon sequestration potential makes agroforestry a critical tool in the global effort to combat climate change.

Additionally, agroforestry systems can help to conserve and enhance biodiversity. By creating diverse and complex habitats, agroforestry supports a wide range of flora and fauna, including pollinators, natural predators, and other beneficial organisms. This increased biodiversity not only contributes to the overall health of ecosystems but also provides essential ecosystem services, such as **pest control** and **pollination**, which are vital for sustainable agriculture.

Agroforestry also plays a role in climate change adaptation. The integration of trees into agricultural systems can help buffer against extreme weather events, such as droughts, floods, and storms. Trees can act as windbreaks, protecting crops from damaging winds, and their root systems can help to stabilize soils, reducing the risk of landslides and erosion. Moreover, the shade provided by trees can help to moderate temperatures, reducing heat stress on crops and livestock.

Lastly, agroforestry can contribute to the sustainable management of water resources. For example, trees can improve water infiltration and retention in soils, reducing surface runoff and the risk of flooding. They can also help to maintain water quality by filtering pollutants and preventing sedimentation in water bodies. These benefits are particularly important in increasing water scarcity and the need for efficient water use in agriculture.

In conclusion, agroforestry is crucial in promoting sustainable agriculture and mitigating the effects of climate change. By integrating trees into agricultural systems, agroforestry offers a multitude of environmental, social, and economic benefits that contribute to a more resilient and sustainable future. As we look ahead, the continued adoption and innovation of agroforestry and silvopasture practices will be essential in addressing the global challenges of food security and climate change.

The Future of Agroforestry and Silvopasture

As we reach the end of our journey through the history and evolution of agroforestry and silvopasture, it is essential to look ahead and envision the future of these sustainable land management practices. In a world grappling with the consequences of climate change, deforestation, and food insecurity, the importance of agroforestry and silvopasture cannot be overstated. These practices offer a promising path toward a more sustainable and resilient future for agriculture, ecosystems, and global communities.

The future of agroforestry and silvopasture will likely be shaped by several critical factors, including technological advancements, policy support, and increased awareness and adoption by farmers and landowners. As technology continues to advance, we can expect innovations to emerge that will further enhance the efficiency and effectiveness of agroforestry and silvopasture systems. For instance, precision agriculture tools, such as drones and satellite imagery, can help farmers monitor and manage their agroforestry systems more effectively, while advancements in plant breeding and genetics can lead to the development of more productive and resilient tree and crop species.

Policy support will also play a crucial role in the future of agroforestry and silvopasture. Governments and international organizations must recognize the value of these practices in promoting sustainable agriculture, conserving biodiversity, and mitigating climate change. By providing financial incentives, technical assistance, and favorable regulatory frameworks, policymakers can encourage the widespread adoption of agroforestry and silvopasture systems. Additionally, integrating agroforestry and silvopasture into national and international climate change strategies can help countries meet their greenhouse gas emissions reduction targets while enhancing food security and rural livelihoods.

Increased awareness and adoption of agroforestry and silvopasture practices by farmers and landowners is another critical factor for the future of these systems. Education and outreach efforts must be scaled up to ensure that farmers are aware of the benefits of agroforestry and silvopasture and have access to the knowledge and resources needed to implement these practices successfully. This can be achieved through farmer-to-farmer knowledge exchange, capacity-building programs, and the development of demonstration sites showcasing the benefits of agroforestry and silvopasture systems.

In conclusion, the future of agroforestry and silvopasture is bright, with immense potential for transforming agriculture and land management practices worldwide. By embracing these sustainable systems, we can work towards a future where agriculture and forestry coexist harmoniously, providing food, income, and ecosystem services while also mitigating the impacts of climate change. The task ahead is not an easy one, but with continued innovation, policy support, and widespread adoption, agroforestry and silvopasture can play a pivotal role in shaping a more sustainable and resilient world for generations to come.

Chapter Summary

- Agroforestry, the integration of trees and shrubs into agricultural systems, has been practiced by various cultures across the globe since the dawn of agriculture, providing numerous benefits such as improved soil fertility, enhanced biodiversity, and increased resilience against climate change.
- Ancient civilizations, including the indigenous cultures of the Amazon Basin, ancient African societies, the Mayans, and traditional rice paddy systems in Asia, all utilized agroforestry techniques to sustain their communities and foster environmental stewardship.
- Silvopasture, a practice that combines forestry and the grazing of domesticated animals, has a rich and diverse

history, with roots in ancient civilizations and evolving over time to suit different environments and cultures.
- Modern agroforestry has experienced a resurgence due to the need for innovative solutions to address pressing global challenges, such as food security, climate change, and environmental degradation.
- Innovations in modern agroforestry include the development of new, highly productive tree species and cultivars, advanced spatial planning and design tools, and a growing body of scientific research that quantifies the ecological, economic, and social benefits of agroforestry practices.
- Challenges faced in implementing agroforestry practices include a lack of awareness and understanding among farmers, policymakers, and the general public, the need for supportive policies and incentives, and the need for continued research and innovation to address specific challenges in different regions and contexts.
- Agroforestry is crucial in promoting sustainable agriculture and mitigating the effects of climate change by enhancing productivity and resilience, sequestering carbon, conserving and enhancing biodiversity, and contributing to climate change adaptation and sustainable water resource management.
- The future of agroforestry and silvopasture will be shaped by technological advancements, policy support, and increased awareness and adoption by farmers and landowners, with the potential to transform agriculture and land management practices worldwide and contribute to a more sustainable and resilient future.

2

PRINCIPLES AND CONCEPTS OF AGROFORESTRY

An image that showcases the symbiotic relationship between trees, crops, and livestock in a thriving agroforestry system, highlighting the interdependence and harmony between these elements.

Agroforestry, an age-old practice that has been refined over generations, is a sustainable land management system that integrates trees, crops, and livestock on the same piece of land. This harmonious coexistence of diverse species optimizes the use of natural resources and offers many ecological, social, and economic benefits. In this chapter, we will delve into the principles and concepts that underpin agroforestry, providing a solid foundation for understanding the intricacies of this fascinating approach to land management.

At the heart of agroforestry lies the idea of **working with nature** rather than against it. By mimicking the structure and function of natural ecosystems, agroforestry systems are designed to enhance productivity, resilience, and sustainability. This is achieved through carefully selecting and arranging tree, crop, and livestock components, which are chosen based on their complementary traits and interactions. The result is a dynamic, self-regulating system that provides multiple benefits while minimizing the need for external inputs and reducing the risk of environmental degradation.

One of the key principles of agroforestry is **diversity.** By incorporating a wide range of plant and animal species, agroforestry systems can capitalize on the unique strengths and characteristics of each component while also buffering against potential threats such as pests, diseases, and climate variability. This diversity also promotes beneficial interactions among species, such as nutrient cycling, pollination, and natural pest control, which contribute to the overall health and productivity of the system.

Another important concept in agroforestry is **spatial and temporal integration.** This refers to the strategic arrangement of trees, crops, and livestock in space and time to maximize their complementary interactions and minimize competition for resources such as light, water, and nutrients. For example, trees can be planted in rows or clusters to provide shade and wind protection for crops while allowing enough sunlight to penetrate the canopy for optimal growth. Similarly, livestock can be rotated through different parts of the agroforestry system to help control weeds, fertilize the soil, and reduce the buildup of pests and diseases.

In addition to these ecological principles, agroforestry encompasses various social and economic considerations. For instance, agroforestry systems must be designed to meet the needs and preferences of local communities, taking into account factors such as land tenure, labor availability, and market opportunities. By providing a diverse array of products and services, agroforestry can help to enhance food security, generate income, and improve the livelihoods of rural households while also contributing to broader goals such as climate change mitigation and biodiversity conservation.

In the following sections, we will explore the ecological foundation of agroforestry systems, the key components and design considerations involved, and the role of agroforestry in sustainable land management. We will also examine the socioeconomic benefits and challenges associated with agroforestry adoption and discuss the prospects of agroforestry and silvopasture in the face of global environmental change.

The Ecological Foundation of Agroforestry Systems

The ecological foundation of agroforestry systems is rooted in the harmonious integration of trees, crops, and livestock, which work together to create a sustainable and productive landscape. By mimicking natural ecosystems, agroforestry systems promote biodiversity, enhance soil health, and contribute to climate change mitigation. In this section, we will delve into the ecological principles that underpin agroforestry systems and explore how these principles contribute to the overall sustainability and resilience of the landscape.

One of the core ecological principles of agroforestry is the concept of **diversity**. In nature, diverse ecosystems are more resilient and productive than monocultures. Agroforestry systems capitalize on this principle by integrating a variety of plant and animal species, each with its unique roles and functions. This diversity not only enhances the system's overall productivity but also reduces the risk of pests and diseases, as well as the reliance on external inputs such as fertilizers and pesticides.

Another fundamental ecological principle in agroforestry is the **efficient use of resources.** By integrating trees, crops, and livestock, agroforestry systems can optimize the use of available resources such as sunlight, water, and nutrients. For example, trees can provide shade and shelter for crops and livestock, reducing the need for irrigation and protecting against extreme weather events. In addition, trees and crops can be strategically planted to improve nutrient cycling and reduce soil erosion, leading to healthier and more fertile soils.

The ecological foundation of agroforestry systems also includes the principle of **mutualism** and **symbiosis.** In these systems, different species interact in ways that benefit one another, creating a more resilient and productive landscape. For instance, nitrogen-fixing trees can improve soil fertility by converting atmospheric nitrogen into a form that crops can utilize. Similarly, livestock can contribute to nutrient cycling and soil fertility by grazing on crop residues and depositing manure.

Agroforestry systems are also crucial in **carbon sequestration** and **climate change mitigation.** Trees and other perennial plants in these systems absorb and store atmospheric carbon dioxide, helping to reduce greenhouse gas emissions. Moreover, agroforestry practices can enhance the overall resilience of the landscape to climate change by promoting biodiversity, improving water and nutrient cycling, and reducing the risk of soil erosion.

In summary, the ecological foundation of agroforestry systems is built upon the principles of diversity, resource efficiency, mutualism, and climate change mitigation. By integrating trees, crops, and livestock in a manner that mimics natural ecosystems, agroforestry systems contribute to the overall sustainability and resilience of the landscape. In the following sections, we will explore the key components and design considerations of agroforestry systems and their role in sustainable land management and socioeconomic development.

Key Components and Design Considerations in Agroforestry

The success of an agroforestry system depends on the careful selection and integration of its key components, which include trees, crops, and livestock. These components must be chosen and arranged to optimize the system's ecological and socioeconomic benefits. This section will discuss the essential elements of agroforestry and the design considerations that must be considered when planning and implementing an agroforestry system.

Trees: Trees are the backbone of any agroforestry system, providing many ecological and economic benefits. When selecting tree species for an agroforestry system, it is crucial to consider factors such as growth rate, canopy structure, root system, and the ability to fix nitrogen. Additionally, the choice of tree species should be based on their compatibility with the local climate, soil conditions, and the specific needs of the crops and livestock in the system.

Crops: Crops are an integral part of agroforestry systems, providing food, fodder, and income for farmers. The selection of crops should be based on their adaptability to the local climate and soil conditions and their compatibility with the chosen tree species. When selecting crops for an agroforestry system, it is essential to consider factors such as light requirements, water needs, and nutrient demands.

Livestock: Incorporating livestock into an agroforestry system can enhance its productivity and sustainability by providing additional income sources, improving nutrient cycling, and promoting natural pest control. The choice of livestock species should be based on their compatibility with the tree and crop components and their ability to thrive in the local environment.

Design Considerations

When designing an agroforestry system, several factors must be considered to ensure its success. These include:

- **Spatial arrangement:** The layout of trees, crops, and livestock should be planned to optimize the use of space, light, and resources while minimizing competition between components.
- **Integration of components:** The interactions between trees, crops, and livestock should be carefully considered to promote synergies and minimize negative impacts. For example, selecting trees that provide shade and shelter for crops and livestock can enhance their productivity and well-being.
- **Management practices:** The choice of management practices, such as pruning, fertilization, and pest control, should be tailored to the specific needs of the agroforestry system and its components.
- **Market considerations:** The choice of tree, crop, and livestock species should be based on their market potential and the needs of the local community, ensuring that the system provides a reliable source of income for farmers.

In conclusion, the key components and design considerations in agroforestry systems are crucial for their success and sustainability. By carefully selecting and integrating trees, crops, and livestock and considering factors such as spatial arrangement, management practices, and market potential, agroforestry systems can provide various ecological and socioeconomic benefits for farmers and the environment.

The Role of Agroforestry in Sustainable Land Management

In today's world, sustainable land management has become crucial to preserving our environment and ensuring food security for future generations. Agroforestry, an integrative approach that combines agricultural and forestry practices, plays a vital role in achieving this goal. This section will explore how agroforestry contributes to sustainable land management through its various ecological, economic, and social benefits.

One of the primary ways agroforestry promotes sustainable land management is through its ability to **enhance soil health.** Integrating trees and shrubs in agricultural systems helps improve soil fertility by fixing atmospheric nitrogen, recycling nutrients, and increasing organic matter content. This, in turn, leads to improved soil structure, water retention, and reduced erosion. Moreover, the deep-rooting nature of trees helps to prevent soil compaction and maintain a healthy soil profile, which is essential for the long-term productivity of the land.

Agroforestry systems also contribute to sustainable land management by promoting **biodiversity.** Integrating various tree species and crops creates a diverse and dynamic ecosystem that supports a wide range of flora and fauna. This increased biodiversity enhances the system's resilience against pests and diseases and provides essential habitat for pollinators and other beneficial organisms. Furthermore, agroforestry can serve as a valuable tool for landscape connectivity, creating corridors for wildlife movement and promoting the conservation of native species.

Another significant aspect of agroforestry's role in sustainable land management is its potential for **climate change mitigation** and adaptation. Trees in agroforestry systems act as carbon sinks, sequestering atmospheric carbon dioxide and reducing greenhouse gas emissions. Additionally, the increased tree cover provides shade and microclimatic benefits, helping to regulate temperature and humidity within the system. This can be particularly beneficial in the face of increasing temperatures and erratic rainfall patterns associated with climate change. Agroforestry systems also contribute to climate change adaptation by enhancing the resilience of agricultural landscapes to extreme weather events, such as floods and droughts, through improved soil health and water management.

From an economic perspective, agroforestry can contribute to sustainable land management by diversifying farm income and reducing the risks associated with monoculture systems. The integration of trees and crops allows farmers to tap into multiple markets and income streams, providing a safety net during crop failure or fluctuating market prices. Furthermore, agroforestry products, such as timber, fruits, and nuts, often fetch higher market prices than traditional agricultural commodities, contributing to increased farm profitability and long-term viability.

Lastly, agroforestry plays a crucial role in sustainable land management by addressing various social and cultural aspects of rural communities. The adoption of agroforestry practices can lead to improved **food security and nutrition**, as well as increased access to fuelwood, fodder, and other essential resources. Moreover, agroforestry systems can contribute to preserving traditional knowledge and cultural practices, fostering a sense of pride and identity among local communities.

In conclusion, agroforestry is a powerful tool for sustainable land management, offering a range of ecological, economic, and social benefits. By integrating trees and crops in a unified and dynamic system, agroforestry can help to preserve our environment, enhance food security, and promote the well-being of rural communities. As we look towards the future, the continued adoption and development of agroforestry and silvopasture practices will be essential in addressing the complex challenges of climate change, biodiversity loss, and global food security.

Socioeconomic Benefits and Challenges of Agroforestry Adoption

As an innovative and sustainable land management practice, agroforestry offers myriad socioeconomic benefits to farmers, communities, and nations. However, it also presents certain challenges that must be addressed for successful adoption and implementation. This section will explore the various socioeconomic benefits and challenges of agroforestry adoption.

One of the most significant benefits of agroforestry is its potential to **diversify** and **increase farmers' income**. By integrating trees and crops, farmers can generate multiple sources of revenue from the same piece of land. This diversification provides financial stability and reduces the risks associated with monoculture farming, such as crop failure due to pests, diseases, or adverse weather conditions. Additionally, agroforestry systems can provide valuable non-timber forest products (NTFPs), such as fruits, nuts, medicinal plants, and fodder, which farming households can sell or consume.

Agroforestry also contributes to **food security** by enhancing the productivity and resilience of agricultural systems. The integration of trees in farming systems can improve soil fertility, reduce soil erosion, and provide shade and windbreaks, which in turn can increase crop yields. Moreover, agroforestry can help maintain and enhance biodiversity, which is crucial for the long-term sustainability of agriculture and food production.

Another important socioeconomic benefit of agroforestry is its potential to create **employment opportunities**, particularly in rural areas. The labor-intensive nature of agroforestry practices, such as planting, pruning, and harvesting, can generate jobs for local communities. Additionally, the processing and marketing of agroforestry products can stimulate the growth of small and medium enterprises, further contributing to rural development and poverty alleviation.

Despite these numerous benefits, the adoption of agroforestry also presents some challenges. One of the main challenges is the need for more knowledge and technical expertise among farmers and extension agents. The successful implementation of agroforestry systems requires a deep understanding of the ecological and agronomic principles involved and the ability to select appropriate tree species and management practices. This knowledge gap can be addressed through capacity building, training, and the dissemination of agroforestry technologies and best practices.

Another challenge is the relatively long-term nature of agroforestry investments. Trees usually take several years to mature and provide significant economic returns, which may discourage farmers from adopting agroforestry practices. To overcome this barrier, we must promote short-term benefits, such as the provision of fodder, firewood, or NTFPs, and develop financial mechanisms that support long-term investments in agroforestry.

Lastly, the lack of supportive policies and institutional frameworks can hinder the widespread adoption of agroforestry. Therefore, governments and policymakers need to recognize the value of agroforestry in addressing various socioeconomic and environmental challenges and create enabling conditions for its promotion and implementation. This may include the development of agroforestry-specific policies, the integration of agroforestry into existing agricultural and forestry policies, and the provision of incentives for farmers to adopt agroforestry practices.

In conclusion, agroforestry offers a multitude of socioeconomic benefits, including income diversification, food security, and rural development. However, to fully harness its potential, it is crucial to address the challenges related to knowledge, long-term investments, and policy support. By doing so, agroforestry can play a vital role in promoting sustainable land management and improving the livelihoods of millions of people around the world.

The Future of Agroforestry and Silvopasture

As explored throughout this chapter, agroforestry and silvopasture offer many ecological, economic, and social benefits that can contribute to a more sustainable and resilient future for agriculture and land management. By integrating trees, crops, and livestock, these systems can transform how we produce food, fiber, and fuel while providing essential ecosystem services and fostering biodiversity.

The future of agroforestry and silvopasture is bright. More and more farmers, landowners, and policymakers recognize the value of these practices in addressing pressing global challenges such as climate change, food security, and rural development. As we move forward, it is essential to continue advancing our understanding of these systems and refining their design and management to maximize their potential benefits.

One crucial aspect of this endeavor is the ongoing development and dissemination of agroforestry knowledge and best practices. This includes not only scientific research and technical guidance but also the sharing of traditional and indigenous knowledge that has been honed over generations of working with the land. By fostering a global community of agroforestry practitioners and researchers, we can ensure that these systems continue to evolve and adapt to the changing needs of our world.

Another critical factor in the future of agroforestry and silvopasture is the development of supportive policies and incentives at local, national, and international levels. This includes measures such as financial support for agroforestry adoption, the integration of agroforestry into agricultural and environmental policies, and the recognition of agroforestry's contributions to climate change mitigation and adaptation efforts. By creating an enabling environment for agroforestry, we can accelerate its adoption and maximize its impact.

Finally, the future of agroforestry and silvopasture will be shaped by the choices and actions of individual farmers, landowners, and communities. By embracing these systems and incorporating them into their land management strategies, these stakeholders can help to drive a shift towards more sustainable and resilient agricultural landscapes. This, in turn, will contribute to improved livelihoods, enhanced food security, and a healthier planet for all.

In conclusion, agroforestry and silvopasture represent a promising pathway toward a more sustainable and resilient future for agriculture and land management. By integrating trees, crops, and livestock, these systems can help address some of our most pressing challenges while providing a wide range of ecological, economic, and social benefits. As we continue to advance our understanding and practice of agroforestry, we can look forward to a future in which these systems play an increasingly important role in shaping the health and well-being of our planet and its inhabitants.

Chapter Summary

- Agroforestry is a sustainable land management system that integrates trees, crops, and livestock on the same piece of land, offering ecological, social, and economic benefits.

- The principles of agroforestry include diversity, spatial and temporal integration, and working with nature to mimic natural ecosystems, enhancing productivity, resilience, and sustainability.
- Agroforestry systems promote biodiversity, enhance soil health, and contribute to climate change mitigation by sequestering carbon dioxide and improving landscape resilience.
- Key components of agroforestry systems include trees, crops, and livestock, which must be carefully selected and integrated based on their complementary traits and interactions.
- Agroforestry is vital in sustainable land management by improving soil fertility, promoting biodiversity, and mitigating climate change impacts.
- Socioeconomic benefits of agroforestry include income diversification, food security, rural development, and employment opportunities.
- Challenges to agroforestry adoption include a need for knowledge and technical expertise, long-term investment nature, and insufficient supportive policies and institutional frameworks.
- The future of agroforestry and silvopasture is promising, with continued development and dissemination of knowledge, supportive policies, and individual actions driving a shift towards more sustainable and resilient agricultural landscapes.

3

SILVOPASTURE: INTEGRATING LIVESTOCK, TREES, AND FORAGE

A harmonious silvopasture landscape, where livestock, trees, and forage coexist to create a sustainable and productive ecosystem.

In today's rapidly changing world, the need for sustainable agricultural practices has never been more critical. As the global population continues to grow and climate change threatens the stability of our ecosystems, innovative solutions are required to ensure food security while minimizing environmental impacts. One such solution is silvopasture, a sustainable approach to agriculture that integrates livestock, trees, and forage into a single, harmonious system.

Silvopasture is not a new concept; in fact, it has been practiced for centuries by farmers and indigenous communities around the world. However, it is only recently that this ancient practice has gained renewed attention as a viable and sustainable alternative to conventional agriculture. By combining the best of both worlds – the productive capacity of livestock farming and the ecological benefits of forestry – silvopasture offers a unique and promising solution to some of the most pressing challenges facing modern agriculture.

At its core, silvopasture is about **creating a diverse and resilient agricultural landscape** that supports multiple functions and provides a wide range of benefits. This is achieved by carefully selecting and managing the right combination of tree species, forage plants, and livestock to create a productive and self-sustaining system. When done correctly, silvopasture can improve soil health, increase biodiversity, enhance animal welfare, and even boost farm profitability.

In this chapter, we will explore the many benefits of silvopasture, from its environmental advantages to its economic and social contributions. We will delve into the key components of a successful silvopasture system and provide practical guidance on designing and implementing this innovative approach to agriculture. Through real-world case studies, we will demonstrate the transformative potential of silvopasture in action, showcasing its ability to create more resilient and sustainable agricultural systems.

As we embark on this journey into the world of silvopasture, we hope that you will come away with a deeper understanding of this sustainable approach to agriculture and be inspired to consider its potential in your farming endeavors. The future of agriculture depends on our ability to adapt and innovate, and silvopasture offers a promising path forward for a more sustainable and resilient food system.

Benefits of Silvopasture: Environmental, Economic, and Social Advantages

Silvopasture, a sustainable agricultural practice that combines livestock, trees, and forage, offers a myriad of benefits that extend beyond the farm. By integrating these elements, silvopasture systems create a harmonious balance that benefits the environment, the economy, and society as a whole. This section will delve into the numerous advantages of adopting silvopasture practices.

Environmental Benefits

One of the most significant benefits of silvopasture is its positive impact on the environment. By incorporating trees into livestock systems, silvopasture helps combat climate change through carbon sequestration. Trees absorb carbon dioxide from the atmosphere and store it in their biomass, effectively reducing greenhouse gas emissions.

Additionally, silvopasture systems promote biodiversity by providing a diverse habitat for various flora and fauna. Trees and shrubs create a more complex ecosystem, attracting pollinators, birds, and other wildlife. This increased biodiversity contributes to more robust and resilient ecosystems, which can better withstand the impacts of climate change and other environmental stressors.

Silvopasture also helps to conserve water resources. Tree roots help to retain water in the soil, reducing runoff and erosion. This conserves water and improves water quality by filtering out pollutants and sediment.

Economic Benefits

Silvopasture systems offer numerous economic advantages for

farmers and landowners. By diversifying farm income, silvopasture can provide a more stable financial foundation. Livestock, timber, and forage production can all generate income, reducing reliance on a single commodity.

Furthermore, the integration of trees and livestock can lead to increased productivity. Trees provide shade and shelter for livestock, reducing heat stress and improving animal welfare. This, in turn, can lead to higher weight gains and improved reproductive performance, ultimately increasing the farm's overall profitability.

Silvopasture can also reduce input costs for farmers. The presence of trees and shrubs can reduce the need for supplemental feed, as they provide additional forage for livestock. Moreover, trees can act as natural windbreaks, reducing the need for expensive infrastructure to protect livestock from harsh weather conditions.

Social Benefits

Silvopasture systems contribute to the social well-being of rural communities by promoting food security and preserving cultural heritage. In addition, by diversifying agricultural production, silvopasture can help to ensure a stable food supply, even in the face of climate change and other challenges.

Additionally, silvopasture practices often align with traditional land management techniques, allowing farmers to maintain their **cultural identity** while embracing sustainable agriculture. This can foster a sense of pride and connection to the land, strengthening rural communities.

In conclusion, silvopasture offers many environmental, economic, and social benefits, making it an attractive option for farmers and landowners seeking sustainable agricultural practices. By integrating livestock, trees, and forage, silvopasture systems create a harmonious balance that benefits the farm, the broader environment, and society.

Key Components of a Successful Silvopasture System

Silvopasture is a dynamic and innovative approach to agriculture that combines the best of both worlds: the productive capacity of livestock farming and the ecological benefits of tree cultivation. By integrating these elements, silvopasture systems can create a sustainable and resilient agricultural landscape that provides multiple benefits for farmers, livestock, and the environment. In this section, we will discuss the key components that contribute to the success of a silvopasture system, ensuring its long-term viability and effectiveness.

Careful Selection of Tree Species: The choice of tree species is a crucial factor in the success of a silvopasture system. Trees should be selected based on their compatibility with the local climate, soil conditions, and the livestock's specific needs. Ideally, the chosen tree species should provide shade, shelter, and forage for the animals while contributing to soil fertility and carbon sequestration.

Strategic Tree Planting and Spacing: The arrangement and density of trees in a silvopasture system can significantly impact its overall productivity and functionality. Trees should be planted in a way that maximizes the available sunlight for both the trees and the understory forage while also providing adequate space for livestock to graze and move freely. Proper tree spacing can also help prevent soil compaction and erosion and promote healthy root development and nutrient cycling.

Diverse and Nutritious Forage: A successful silvopasture system relies on a diverse and nutritious forage base to support the health and productivity of the livestock. This can include a mix of grasses, legumes, and other herbaceous plants that provide a balanced diet for the animals and a habitat for beneficial insects and other wildlife. Regular monitoring and management of the forage are essential to ensure its quality and availability throughout the grazing season.

Adaptive Livestock Management: Integrating livestock into a silvopasture system requires careful planning and adaptive management to ensure the animals' well-being and the system's long-term sustainability. This includes selecting appropriate livestock species and breeds, implementing rotational grazing practices, and monitoring animal health and performance. By adapting management practices to the specific needs of the silvopasture system, farmers can optimize livestock productivity while minimizing negative impacts on the environment.

Ongoing Monitoring and Evaluation: A successful silvopasture system is one that evolves and adapts over time, responding to changing conditions and the needs of the farm. Regular monitoring and evaluation of the system's performance can help identify areas for improvement and inform future management decisions. This may involve tracking tree growth and health, forage quality and availability, livestock performance, and other indicators of system health and productivity.

In conclusion, the key components of a successful silvopasture system include careful selection of tree species, strategic tree planting and spacing, diverse and nutritious forage, adaptive livestock management, and ongoing monitoring and evaluation. By incorporating these elements into the design and management of a silvopasture system, farmers can create a resilient and sustainable agricultural landscape that benefits both people and the environment.

Designing and Implementing Silvopasture: Best Practices and Strategies

Silvopasture, a sustainable approach to agriculture, requires careful planning and implementation to ensure its success. By integrating livestock, trees, and forage, farmers can create a harmonious and productive ecosystem that benefits the environment, the economy, and society. This section will explore best practices and strategies for designing and implementing a successful silvopasture system.

Assessing the Site: Before embarking on a silvopasture project, conducting a thorough site assessment is crucial. This includes evaluating the soil type, topography, climate, and existing vegetation. Understanding these factors will help you determine the most suitable tree species, livestock, and forage crops for your system.

Choosing the Right Tree Species: The selection of tree species is a critical factor in the success of a silvopasture system. Consider native species that are well-adapted to the local climate and soil conditions. Additionally, choose trees that provide multiple benefits, such as shade, timber, and fodder for livestock.

Integrating Livestock: The choice of livestock should be based on the available resources, market demand, and the compatibility of the animals with the chosen tree species. It is essential to ensure that the livestock's grazing habits do not negatively impact the growth and development of the trees.

Forage Management: A successful silvopasture system requires careful management of the forage component. This includes selecting appropriate forage species, implementing rotational grazing practices, and monitoring the health and productivity of the forage crops.

Tree Protection and Maintenance: Young trees in a silvopasture system are vulnerable to damage from livestock. Employ protective measures, such as tree guards or temporary fencing, to prevent browsing and trampling. Regular pruning and thinning of the trees will also help to maintain their health and productivity.

Monitoring and Evaluation: Regular monitoring of the silvopasture system is essential to ensure its success. Keep track of tree growth, forage production, and livestock health to identify any issues that may arise. Adjust management practices as needed to maintain the system's productivity and sustainability.

Education and Training: Silvopasture is a complex and dynamic system that requires specialized knowledge and skills. Invest in education and training for yourself and your team to ensure that you are equipped to manage the system effectively.

By following these best practices and strategies, you can design and implement a successful silvopasture system that provides various environmental, economic, and social benefits. As more farmers adopt this sustainable approach to agriculture, we can look forward to a future where our agricultural landscapes are more resilient, diverse, and productive.

Case Studies: Real-World Examples of Silvopasture in Action

Silvopasture systems have been successfully implemented in various regions around the world, showcasing the adaptability and effectiveness of this sustainable agricultural practice. In this section, we will explore several real-world examples of silvopasture in action, highlighting the unique challenges and successes experienced by farmers and land managers who have embraced this innovative approach to agriculture.

Case Study 1: Pine Forest Silvopasture in the Southeastern United States

In the southeastern United States, a growing number of farmers are integrating livestock, trees, and forage crops in a silvopasture system based on pine forests. This practice involves thinning existing pine stands to allow for the growth of forage crops, such as grasses and legumes, which provide nutritious feed for grazing animals like cattle and sheep. The trees offer shade and shelter for the livestock, while their roots help to prevent soil erosion and improve nutrient cycling. Farmers in this region have reported increased livestock productivity, improved timber quality, and enhanced wildlife habitat and carbon sequestration benefits.

Case Study 2: Coffee Agroforestry in Central America

In Central America, coffee farmers have been integrating shade trees and livestock into their coffee plantations to create a more resilient and sustainable agroecosystem. By incorporating trees and shrubs that provide shade, farmers can reduce the need for chemical inputs, protect their crops from pests and diseases, and improve soil fertility. Livestock, such as chickens and pigs, can be raised alongside the coffee plants, providing additional income and contributing to the farm's overall productivity. This silvopasture system has been shown to increase coffee yields, improve the quality of the beans, and enhance biodiversity.

Case Study 3: Temperate Silvopasture in the United Kingdom

In the United Kingdom, some farmers are experimenting with silvopasture systems in temperate climates, combining livestock grazing with the cultivation of fruit and nut trees. This approach allows farmers to diversify their income streams and reduce reliance on monoculture cropping systems. By integrating trees and livestock, farmers can improve soil fertility, reduce erosion, and create habitats for beneficial insects and wildlife. The trees also provide valuable timber and non-timber forest products, such as fruits, nuts, and firewood, which can be sold or used on the farm.

Case Study 4: Tropical Silvopasture in Brazil

In Brazil, silvopasture systems have been used to restore degraded pasturelands and improve the productivity of cattle ranching operations. By planting native tree species alongside forage grasses, ranchers can create a more diverse and resilient landscape that supports both livestock and wildlife. The trees provide shade, reducing heat stress for the animals, and their deep roots help to improve soil structure and water infiltration. This approach has been shown to increase livestock weight gain and overall productivity while providing essential ecosystem services, such as carbon sequestration and habitat for native species.

These case studies demonstrate the versatility and effectiveness of silvopasture systems in a variety of contexts and climates. By integrating livestock, trees, and forage crops, farmers and land managers can create more sustainable, productive, and resilient agricultural systems that benefit both people and the environment. As the world faces increasing challenges related to climate change, resource scarcity, and food security, silvopasture offers a promising solution for the future of agriculture.

The Future of Silvopasture and Its Role in Sustainable Agriculture

In conclusion, silvopasture is a promising and innovative approach to sustainable agriculture, offering many environmental, economic, and social benefits. By integrating livestock, trees, and forage into a single, harmonious system, silvopasture can revolutionize how we approach farming and land management. As we face increasing challenges from climate change, population growth, and resource scarcity, the need for sustainable agricultural practices has never been more critical.

The future of silvopasture is bright; more farmers and landowners recognize its advantages and adopt this practice on their lands. Through continued research, education, and collaboration, we can refine and optimize silvopasture systems to maximize their benefits and minimize potential drawbacks. Furthermore, by **sharing knowledge and experiences**, we can help ensure that silvopasture becomes an integral part of sustainable agriculture worldwide.

As we have explored in this chapter, the key components of a successful silvopasture system include careful planning, design, implementation, ongoing management, and monitoring. By adhering to best practices and strategies, farmers can create thriving silvopasture systems that benefit their livestock, trees, forage, the environment, and their bottom line.

The case studies presented in this chapter demonstrate the versatility and adaptability of silvopasture, as it can be applied to various climates, landscapes, and farming systems. These real-world examples inspire and guide those considering the implementation of silvopasture on their lands.

Ultimately, the widespread adoption of silvopasture and other agroforestry practices can significantly address global challenges such as climate change, biodiversity loss, and food security. Furthermore, by embracing silvopasture, we can work towards a more sustainable, resilient, and productive agricultural future for generations to come.

Chapter Summary

- Silvopasture is a sustainable agricultural practice that integrates livestock, trees, and forage into a single system, offering environmental, economic, and social benefits.
- Environmental benefits of silvopasture include carbon sequestration, increased biodiversity, and improved water conservation and quality.
- Economic benefits of silvopasture include diversified farm income, increased productivity, and reduced input costs for farmers.
- Social benefits of silvopasture include promoting food security and preserving cultural heritage in rural communities.
- Key components of a successful silvopasture system include careful selection of tree species, strategic tree planting and spacing, diverse and nutritious forage, adaptive livestock management, and ongoing monitoring and evaluation.
- Best practices for designing and implementing silvopasture include assessing the site, choosing the right tree species, integrating livestock, managing forage, protecting and maintaining trees, monitoring and evaluating the system, and investing in education and training.
- Real-world case studies demonstrate the versatility and effectiveness of silvopasture systems in various regions and climates, showcasing their potential to create more sustainable, productive, and resilient agricultural systems.
- The future of silvopasture is promising, as its widespread adoption can play a significant role in addressing global challenges such as climate change, biodiversity loss, and food security, contributing to a more sustainable and resilient agricultural future.

4

AGROFORESTRY SYSTEMS AND PRACTICES

A thriving agroforestry system, where diverse crops and trees are strategically integrated, creating a harmonious and sustainable agricultural landscape.

Agroforestry, a term coined in the *1970s*, is an age-old practice revitalized in recent years as a sustainable approach to agriculture and land management. At its core, agroforestry is the integration of trees, crops, and livestock into a single, ecologically balanced system. This harmonious coexistence of various elements not only enhances the productivity of the land but also provides numerous environmental, social, and economic benefits.

This chapter will explore the fascinating world of agroforestry systems and practices, exploring traditional and modern approaches. We will examine agroforestry's key components and benefits, with a particular focus on silvopasture – a unique practice that combines forestry and livestock management. Through case studies, we will showcase successful implementations of agroforestry and silvopasture around the world, and finally, we will discuss the prospects of these sustainable land management techniques.

As we embark on this journey, we must understand that agroforestry is not a one-size-fits-all solution. Instead, it is a highly adaptable practice that can be tailored to suit the specific needs and conditions of a given region, climate, and community. The beauty of agroforestry lies in its flexibility and its ability to foster a symbiotic relationship between humans and nature, ensuring the long-term health and prosperity of both.

So, let's begin our exploration of agroforestry systems and practices and discover how this ancient wisdom can help shape a more sustainable and resilient future for our planet and its inhabitants.

Traditional and Modern Agroforestry Systems

Agroforestry, the practice of integrating trees and shrubs with crops and livestock, has been a part of human agriculture for centuries. Traditional and modern agroforestry systems have evolved to meet the needs of diverse communities and environments, offering a range of benefits for both people and the planet. This section will explore the key characteristics and examples of traditional and modern agroforestry systems, highlighting their unique features and contributions to sustainable agriculture.

Traditional agroforestry systems have been shaped by generations of farmers who have skillfully combined their knowledge of local ecosystems with their agricultural needs. These systems are often characterized by a high degree of biodiversity, as farmers have learned to cultivate various plant and animal species that complement and support one another. Some examples of traditional agroforestry systems include:

Homegardens: These are found in many tropical and subtropical regions. Homegardens are small-scale, multi-layered agroforestry systems that incorporate diverse trees, shrubs, and crops. They often serve as a primary source of food, medicine, and income for rural households while providing valuable ecosystem services such as soil conservation and carbon sequestration.

Shifting cultivation: Also known as swidden agriculture or slash-and-burn, shifting cultivation is a traditional agroforestry practice in which farmers clear a small forest area, burn the vegetation, and then plant crops in the nutrient-rich ash. After a few years of cultivation, the land is left fallow to regenerate, and the process is repeated in a new area. This system allows for the sustainable use of forest resources while maintaining soil fertility and biodiversity.

Forest farming: In this system, farmers cultivate shade-tolerant crops such as coffee, cocoa, or spices under the canopy of a managed forest. This practice provides a sustainable source of income for farmers and helps conserve forest ecosystems and their associated biodiversity.

On the other hand, modern agroforestry systems have emerged in response to the challenges posed by industrial agriculture and the need for more sustainable land management practices. These systems often incorporate scientific knowledge and technological innovations to optimize productivity, resource use efficiency, and environmental benefits. Some examples of modern agroforestry systems include:

Alley cropping: This system involves planting rows of trees or shrubs alongside annual or perennial crops, creating a series of "alleyways" for cultivation. The trees provide shade, wind protection, and nutrient cycling, while the crops benefit from improved microclimates and reduced soil erosion. Common tree species used in alley cropping include nitrogen-fixing species such as Leucaena or Gliricidia, which can also provide fodder for livestock.

Riparian buffer strips: In this system, trees and shrubs are planted along watercourses to protect them from agricultural runoff and erosion. The vegetation acts as a natural filter, absorbing nutrients and sediment while providing wildlife habitat and stabilizing stream banks.

Windbreaks and shelterbelts: These linear plantings of trees and shrubs are designed to protect crops, livestock, and infrastructure from wind damage, reduce soil erosion, and conserve energy. Windbreaks can also enhance biodiversity by providing habitat for birds and other wildlife.

In conclusion, traditional and modern agroforestry systems offer diverse practices that can be adapted to suit different environmental and socio-economic contexts. Furthermore, by integrating trees and shrubs with crops and livestock, these systems contribute to sustainable agriculture, food security, and environmental conservation, making them essential to our efforts to build a more resilient and sustainable future.

Key Components and Benefits of Agroforestry

As an innovative and sustainable land management practice, agroforestry combines the cultivation of trees, crops, and livestock in a mutually beneficial system. This approach optimizes land use and provides numerous environmental, economic, and social benefits. In this section, we will delve into the critical components of agroforestry and explore the advantages it offers over conventional agricultural practices.

The foundation of agroforestry lies in its three primary components: **trees, crops, and livestock**. These elements interact with one another in various ways, creating a dynamic and resilient ecosystem. Trees provide shade, shelter, and habitat for wildlife, while their roots help stabilize the soil and prevent erosion. Crops, in turn, benefit from the improved soil structure and increased organic matter provided by the trees. Livestock can graze on the understory vegetation and contribute to nutrient cycling through their manure. The integration of these components results in a diverse and productive system that can withstand fluctuations in climate and market conditions.

The benefits of agroforestry are numerous and can be categorized into environmental, economic, and social advantages.

Environmental Benefits

1. **Soil Conservation:** Tree roots help stabilize the soil, reducing erosion and improving soil structure. Adding organic matter from leaf litter and root decomposition also enhances soil fertility.
2. **Biodiversity:** Agroforestry systems support a wide range of plant and animal species, promoting ecological balance and preserving habitats for beneficial insects, birds, and mammals.
3. **Carbon Sequestration:** Trees absorb and store carbon dioxide, helping to mitigate climate change. The integration of trees in agricultural landscapes can significantly increase the overall carbon sequestration potential of the land.
4. **Water Management:** Agroforestry practices can improve water infiltration and retention, reducing the risk of flooding and drought. Trees also act as natural filters, improving water quality by absorbing pollutants and nutrients.

Economic Benefits:

1. **Diversified Income:** Agroforestry systems offer multiple sources of income, including timber, non-timber forest

products, and livestock. This diversification reduces the risk associated with market fluctuations and crop failures.
2. **Increased Productivity:** The complementary nature of agroforestry components can lead to higher overall yields. For example, shade-tolerant crops can benefit from the shelter trees provide, while nitrogen-fixing trees can improve soil fertility for neighboring crops.
3. **Cost Savings:** Agroforestry practices can reduce the need for chemical fertilizers, pesticides, and irrigation, lowering production costs and increasing profit margins.

Social Benefits

1. **Food Security:** The diversification of crops and livestock in agroforestry systems can improve food security and nutrition for local communities.
2. **Cultural Preservation:** Many traditional agroforestry practices are rooted in indigenous knowledge and culture, and their continued use helps preserve these valuable traditions.
3. **Community Development:** Agroforestry can provide opportunities for education, training, and employment, fostering social cohesion and empowering rural communities.

In conclusion, the key components of agroforestry – trees, crops, and livestock – work together to create a sustainable and resilient system that offers a multitude of environmental, economic, and social benefits. By embracing agroforestry practices, farmers and land managers can contribute to a more sustainable and productive future for agriculture and forestry.

Silvopasture: Integrating Livestock and Forestry

Silvopasture, a harmonious marriage of livestock and forestry, is a practice that has been gaining momentum in recent years. As a subcategory of agroforestry, silvopasture combines the cultivation of trees with the rearing of livestock, creating a mutually beneficial ecosystem. This innovative approach to land management not only optimizes the use of available resources but also offers a plethora of environmental, economic, and social benefits.

At the heart of silvopasture lies the concept of **symbiosis.** Trees provide shade, shelter, and forage for livestock, while the animals contribute to the health of the soil through their natural grazing and waste production. This dynamic relationship results in improved animal welfare, increased biodiversity, and enhanced soil fertility, making silvopasture a sustainable and efficient land management practice.

One of the key advantages of silvopasture is its ability to **diversify income streams** for farmers. Farmers can tap into multiple markets by integrating trees and livestock, such as timber, non-timber forest products, and animal products like meat, milk, and wool. This diversification not only increases the overall profitability of the farm but also provides a safety net in case one market faces challenges.

Another compelling aspect of silvopasture is its potential to mitigate climate change. Trees act as carbon sinks, absorbing and storing atmospheric carbon dioxide while providing shade that reduces the need for energy-intensive cooling systems for livestock. Moreover, the integration of trees and livestock can reduce greenhouse gas emissions, as improved soil health and fertility promote better nutrient cycling and reduce the need for synthetic fertilizers.

Silvopasture also plays a vital role in promoting biodiversity. By creating a varied landscape with different plant species, habitats, and microclimates, silvopasture systems encourage the growth of diverse flora and fauna. This increased biodiversity benefits the ecosystem's overall health and provides essential ecosystem services such as pollination, pest control, and nutrient cycling.

Implementing silvopasture, however, is not without its challenges. It requires careful planning, management, and monitoring to maintain the balance between trees and livestock. Farmers must consider factors such as tree species selection, planting density, and grazing patterns to create a productive and sustainable system.

In conclusion, silvopasture is a promising and innovative land management approach with many benefits. By integrating livestock and forestry, this practice optimizes the use of available resources and contributes to environmental conservation, economic stability, and social well-being. As we look toward the future of agroforestry and silvopasture, it is essential to continue exploring and refining these practices to ensure their long-term success and sustainability.

Case Studies: Successful Agroforestry and Silvopasture Implementations

In this section, we will delve into real-life examples of successful agroforestry and silvopasture implementations from around the world. These case studies showcase the potential of these systems to enhance agricultural productivity, promote biodiversity, and contribute to sustainable land management practices.

Case Study 1: The Chagga Home Gardens of Tanzania

The Chagga people of Mount Kilimanjaro in Tanzania have practiced agroforestry for centuries. Their home gardens, known as *"vihamba,"* are an excellent example of traditional agroforestry systems. These gardens consist of a diverse mix of trees, shrubs, and crops, including coffee, bananas, and various fruit trees. The multi-layered canopy structure provides shade for the understory crops, while the trees help to conserve soil moisture and prevent erosion. The Chagga home gardens are not only productive but also support a high level of biodiversity, contributing to the overall resilience of the system.

Case Study 2: The Quesungual System in Honduras

In the dry and degraded hillsides of western Honduras, farmers have developed an innovative agroforestry system known as the Quesungual system. This system involves the selective pruning of native trees to create a sparse canopy, allowing sunlight to reach the understory crops such as maize, beans, and sorghum. The pruned branches are used as mulch, which helps to conserve soil moisture and improve soil fertility. The Quesungual system has been shown to increase crop yields, reduce soil erosion, and enhance biodiversity compared to conventional agricultural practices in the region.

Case Study 3: Silvopasture in the Dehesa of Spain

The Dehesa is a traditional silvopastoral system found in the Iberian Peninsula, characterized by the integration of livestock grazing with oak tree forests. The oak trees provide shade and acorns for the grazing animals, such as the Iberian pig and Merino sheep, while the livestock help to maintain the grasslands and promote tree regeneration. This symbiotic relationship results in a sustainable and productive system supporting diverse flora and fauna. The Dehesa is also known for its high-quality products, such as the famous *Iberian ham*, which are highly valued in local and international markets.

Case Study 4: Alley Cropping in the United States

Alley cropping is an agroforestry practice that involves planting rows of trees or shrubs alongside annual or perennial crops. This system has been successfully implemented in various regions across the United States, providing multiple benefits such as improved soil fertility, erosion control, and increased crop yields. One notable example is a farm in Missouri, where a farmer has integrated walnut trees with soybean and maize crops. The walnut trees provide valuable timber and nuts, while the soybean and maize crops benefit from the nitrogen-fixing properties of the trees and the improved microclimate created by the tree canopy.

In conclusion, these case studies demonstrate the versatility and adaptability of agroforestry and silvopasture systems in different ecological and socio-economic contexts. By learning from these successful implementations, we can continue to promote and develop sustainable land management practices that benefit both people and the environment.

The Future of Agroforestry and Silvopasture

As we reach the end of this enlightening journey through the world of agroforestry and silvopasture, it is essential to reflect on the immense potential these practices hold for the future of agriculture, forestry, and environmental conservation. Integrating trees, crops, and livestock in a single system has proven to be a sustainable and productive approach to land management, offering myriad benefits for farmers, communities, and the environment.

The future of agroforestry and silvopasture is undoubtedly promising. As the global population grows, the demand for food, fiber, and fuel will only increase. Agroforestry systems have the potential to meet these demands while simultaneously addressing the challenges of climate change, biodiversity loss, and soil degradation. By adopting these practices, farmers can diversify their income sources, reduce their vulnerability to market fluctuations, and improve their overall resilience in the face of environmental and economic uncertainties.

Furthermore, the benefits of agroforestry and silvopasture extend beyond the farm gate. These systems can play a crucial role in mitigating climate change by sequestering carbon, reducing greenhouse gas emissions, and promoting renewable energy sources. Additionally, they can contribute to biodiversity conservation by providing a habitat for various species of flora and fauna and maintaining the ecological balance of the landscape.

In order to harness the full potential of agroforestry and silvopasture, it is essential to invest in research, education, and extension services. This will enable farmers to access the knowledge, skills, and resources necessary to implement these systems effectively and efficiently. Moreover, supportive policies and incentives should be implemented to encourage the widespread adoption of agroforestry and silvopasture practices.

In conclusion, agroforestry and silvopasture represent a paradigm shift in the way we approach agriculture and forestry. By embracing these systems, we can cultivate a more sustainable, resilient, and productive future for our planet and its inhabitants. The journey may be long and challenging, but the rewards are well worth the effort. So, as we move forward, let us continue to **explore, innovate, and collaborate** in our quest to create a greener, more prosperous world through the power of agroforestry and silvopasture.

Chapter Summary

- Agroforestry is the integration of trees, crops, and livestock into a single, ecologically balanced system, offering numerous environmental, social, and economic benefits.
- Traditional and modern agroforestry systems have evolved to meet the needs of diverse communities and environments. Examples include homegardens, shifting cultivation, forest farming, alley cropping, riparian buffer strips, and windbreaks.
- The key components of agroforestry are trees, crops, and livestock, which interact with one another to create a dynamic and resilient ecosystem.
- Environmental benefits of agroforestry include soil conservation, biodiversity, carbon sequestration, and water management.
- Economic benefits of agroforestry include diversified income, increased productivity, and cost savings.
- Social benefits of agroforestry include food security, cultural preservation, and community development.
- Silvopasture is a subcategory of agroforestry that combines the cultivation of trees with the rearing of livestock, offering

benefits such as diversification of income streams, climate change mitigation, and promotion of biodiversity.
- The future of agroforestry and silvopasture is promising, with the potential to address global challenges such as food security, climate change, and biodiversity loss, but it requires investment in research, education, and supportive policies to encourage widespread adoption.

5

THE ROLE OF AGROFORESTRY IN CLIMATE CHANGE MITIGATION

An image that showcases the harmonious coexistence of agriculture and forestry in a futuristic landscape, where agroforestry practices play a vital role in combating climate change and preserving biodiversity.

Climate change has emerged as one of the most pressing challenges of our time, with far-reaching consequences for ecosystems, economies, and societies worldwide. As global temperatures continue to rise, the need for innovative and sustainable solutions to mitigate the impacts of climate change becomes increasingly urgent. One such solution lies in the realm of agroforestry, a land management practice that integrates trees and shrubs with crops and livestock systems. This chapter explores the multifaceted role of agroforestry in climate change mitigation, delving into its potential to sequester carbon, enhance biodiversity, promote sustainable land management, and provide socioeconomic benefits to communities.

The interconnectedness of agroforestry and climate change is evident in the numerous ways that these systems can influence one another. On the one hand, climate change can exacerbate existing challenges faced by agricultural systems, such as soil degradation, water scarcity, and reduced crop yields. On the other hand, agroforestry practices can help to address these challenges by enhancing the resilience of ecosystems and providing a range of ecosystem services that contribute to climate change mitigation and adaptation. By integrating trees and shrubs into agricultural landscapes, agroforestry systems can create more diverse and resilient ecosystems that are better equipped to withstand the impacts of a changing climate.

At the heart of agroforestry's potential to mitigate climate change is its **capacity for carbon sequestration**, or the process by which trees and other plants capture and store atmospheric carbon dioxide. As trees grow, they absorb carbon dioxide from the atmosphere and convert it into biomass, effectively removing this potent greenhouse gas from the atmosphere and reducing its contribution to global warming. In addition to carbon sequestration, agroforestry systems can also enhance biodiversity and ecosystem resilience, providing habitats for a wide range of plant and animal species and promoting the overall health and stability of ecosystems.

Sustainable land management practices are another key component of agroforestry's role in climate change mitigation. By adopting practices such as conservation tillage, cover cropping, and agroforestry, farmers can improve soil health, reduce erosion, and increase water infiltration, all of which contribute to the overall resilience of agricultural systems in the face of climate change. Furthermore, these practices can also help to reduce the need for synthetic fertilizers and pesticides, which are significant sources of greenhouse gas emissions.

The socioeconomic benefits of agroforestry in a changing climate should not be overlooked. By diversifying agricultural systems and providing additional sources of income through the sale of timber, non-timber forest products, and ecosystem services, agroforestry can help to improve the livelihoods of rural communities and enhance their capacity to adapt to the impacts of climate change. Moreover, agroforestry can also contribute to food security by providing a diverse array of nutritious food sources, such as fruits, nuts, and edible leaves.

In conclusion, the interconnectedness of agroforestry and climate change underscores the importance of embracing this land management practice as a pathway to a sustainable future. By harnessing the power of agroforestry to sequester carbon, enhance biodiversity, promote sustainable land management, and provide socioeconomic benefits, we can work towards mitigating the impacts of climate change and building more resilient agricultural systems for generations to come.

Carbon Sequestration: A Natural Solution to Greenhouse Gas Emissions

As we delve into the heart of the matter, it is crucial to understand the concept of carbon sequestration and its significance in mitigating climate change. Simply put, carbon sequestration is the process by which carbon dioxide (CO_2) is captured and stored, preventing it from being released into the atmosphere and contributing to global warming.

The beauty of agroforestry lies in its ability to **sequester carbon naturally**, harnessing the power of trees and vegetation to combat greenhouse gas emissions.

Trees are the unsung heroes in the fight against climate change. Through the process of photosynthesis, they absorb CO_2 from the atmosphere and convert it into oxygen and biomass. This biomass, composed of roots, stems, branches, and leaves, stores carbon, effectively removing it from the atmosphere. Agroforestry systems, which integrate trees with crops and livestock, create a synergistic environment that maximizes carbon sequestration potential.

The benefits of carbon sequestration in agroforestry systems extend beyond the mere absorption and storage of CO_2. By incorporating trees into agricultural landscapes, farmers can create microclimates that help regulate temperature and humidity, reducing the need for energy-intensive irrigation and cooling systems. Moreover, the shade trees provide can reduce soil evaporation, conserve water and maintain soil moisture levels.

In addition to the direct impact on carbon sequestration, agroforestry practices indirectly contribute to climate change mitigation. For instance, using nitrogen-fixing trees can improve soil fertility, reducing the need for synthetic fertilizers that emit nitrous oxide, a potent greenhouse gas. Furthermore, agroforestry systems can prevent soil erosion and degradation, preserving the soil's capacity to store carbon.

The potential of agroforestry to sequester carbon is immense. According to the Intergovernmental Panel on Climate Change (IPCC), agroforestry practices can sequester between 0.4 and 1.2 gigatons of CO_2 per year, making it one of the most promising natural climate solutions. However, it is essential to recognize that the effectiveness of carbon sequestration in agroforestry systems depends on various factors, such as tree species, climate, and management practices.

In conclusion, carbon sequestration through agroforestry offers a natural and sustainable solution to mitigate greenhouse gas emissions. By integrating trees into agricultural systems, we can harness their power to absorb and store carbon, regulate microclimates, and improve soil health. As we continue to explore the role of agroforestry in climate change mitigation, it becomes increasingly evident that embracing these practices is beneficial for the environment and the well-being of future generations.

Enhancing Biodiversity and Ecosystem Resilience in Agroforestry Systems

The beauty of agroforestry lies in its ability to create a harmonious balance between agricultural production and environmental conservation. One of the most significant advantages of agroforestry systems is their potential to enhance biodiversity and bolster ecosystem resilience. This section will delve into the ways in which agroforestry contributes to the preservation and promotion of biodiversity, as well as the strengthening of ecosystems in the face of climate change.

Biodiversity, or the variety of life in a particular ecosystem, is crucial to a healthy environment. It ensures the **stability and productivity of ecosystems**, which in turn provide essential services such as food, water, and climate regulation. By their very nature, agroforestry systems promote biodiversity by integrating a diverse range of plant and animal species into agricultural landscapes. This integration creates a mosaic of habitats that support a wide array of flora and fauna, including pollinators, predators, and decomposers, all of which play vital roles in maintaining ecosystem health.

In addition to supporting a diverse range of species, agroforestry systems foster genetic diversity within individual species. This is particularly important for crop plants, as genetic diversity provides the raw material for adaptation and evolution in response to changing environmental conditions. By incorporating a variety of crop species and cultivars into agroforestry systems, farmers can create a more resilient agricultural landscape that is better equipped to cope with the challenges posed by climate change.

Ecosystem resilience, or the capacity of an ecosystem to recover from disturbances and maintain its essential functions, is another critical aspect of environmental health. Agroforestry systems can **enhance ecosystem resilience** in several ways. First, by promoting biodiversity, agroforestry systems create a more robust and adaptable ecosystem that can better withstand disturbances such as pests, diseases, and extreme weather events. The presence of multiple species and functional groups within an agroforestry system can also help buffer against the loss of individual species, ensuring that the ecosystem remains functional and productive.

Second, agroforestry systems can help **maintain and improve soil health**, a key factor in ecosystem resilience. Integrating trees and shrubs into agricultural landscapes can reduce soil erosion, improve soil structure, and increase the availability of essential nutrients. These benefits, in turn, promote the growth of a diverse and healthy soil microbial community, which plays a crucial role in nutrient cycling, decomposition, and overall ecosystem functioning.

Lastly, agroforestry systems can contribute to **landscape-scale resilience** by creating connectivity between fragmented habitats. This connectivity allows for the movement of species between patches of suitable habitat, which is essential for maintaining genetic diversity and ensuring the long-term survival of populations. In this way, agroforestry systems can serve as important stepping stones for wildlife in increasingly fragmented landscapes, helping to maintain the ecological integrity of the broader environment.

In conclusion, integrating agroforestry systems into agricultural landscapes offers a promising pathway for enhancing biodiversity and ecosystem resilience in the face of climate change. By fostering a diverse and interconnected network of species and habitats, agroforestry systems can help create a more robust and adaptable environment better equipped to withstand the challenges of a changing climate. As we continue to grapple with the impacts of climate change on our planet, we must embrace and promote agroforestry as a vital component of a sustainable and resilient future.

Sustainable Land Management Practices for Climate Change Adaptation

As the world grapples with the ever-increasing impacts of climate change, it has become imperative to adopt sustainable land management practices that promote climate change adaptation. With its unique blend of agriculture and forestry, agroforestry offers many solutions to help farmers and communities adapt to the changing climate while maintaining productivity and ecosystem health. In this section, we will delve into some of the sustainable land management practices that can be employed in agroforestry systems to **enhance climate change adaptation.**

First and foremost, the selection of appropriate tree species and crop varieties is crucial in building a resilient agroforestry system. By choosing species that are well-adapted to local conditions and resistant to pests and diseases, farmers can ensure the long-term productivity and stability of their agroforestry systems. Moreover, incorporating a diverse mix of species can further enhance resilience, as it reduces the risk of catastrophic losses due to climate-related events such as droughts, floods, or pest outbreaks.

Another key practice in sustainable land management is the conservation of soil and water resources. Agroforestry systems can help in this regard by reducing soil erosion, improving soil structure, and increasing water infiltration and retention. For instance, the strategic planting of trees and shrubs along contour lines or on slopes can act as a barrier against soil erosion, while the deep root systems of trees can help stabilize the soil and improve its structure. Additionally, the presence of trees can enhance water infiltration and retention, reducing the impacts of droughts and floods on crops and local water resources.

Incorporating agroforestry practices that promote natural pest control is another way to enhance climate change adaptation. By fostering a diverse and balanced ecosystem, agroforestry systems can encourage the presence of beneficial insects and other natural predators, reducing the need for chemical pesticides and enhancing the system's overall resilience. For example, integrating insectary plants or creating a habitat for natural predators such as birds and bats can help keep pest populations in check.

Furthermore, agroforestry systems can be designed to optimize the use of available resources, such as sunlight, water, and nutrients, through the strategic arrangement of trees, crops, and livestock. This can be achieved by employing practices such as **alley cropping**, where rows of trees are interspersed with rows of crops, or by integrating livestock into the system to recycle nutrients and reduce the need for external inputs. By making the most of these resources, agroforestry systems can maintain productivity even under changing climate conditions.

Lastly, engaging local communities and farmers in the planning, implementation, and management of agroforestry systems is essential for their long-term success and sustainability. By involving stakeholders in the decision-making process and providing them with the necessary knowledge and tools, agroforestry can empower communities to take charge of their own climate change adaptation strategies.

In conclusion, sustainable land management practices play a vital role in enhancing the adaptive capacity of agroforestry systems in the face of climate change. By incorporating practices that promote resilience, resource conservation, and community engagement, agroforestry can be a powerful tool in our collective efforts to adapt to a changing climate and secure a sustainable future for generations to come.

The Socioeconomic Benefits of Agroforestry in a Changing Climate

As we delve deeper into the multifaceted world of agroforestry, it becomes increasingly evident that this sustainable land management practice offers far more than just environmental benefits. In a world grappling with the consequences of climate change, agroforestry also presents a myriad of socioeconomic advantages that can help communities adapt and thrive in the face of these challenges. In this section, we will explore the various ways in which agroforestry contributes to the socioeconomic well-being of communities and nations alike.

First and foremost, agroforestry systems provide a diversified source of income for farmers and landowners. By integrating trees and shrubs with traditional crops and livestock, agroforestry allows for the simultaneous production of various marketable goods, such as timber, fruits, nuts, and medicinal plants. This diversification not only reduces the risks associated with monoculture farming but also creates new income-generating opportunities for rural communities. Moreover, the integration of high-value tree crops can significantly increase the overall profitability of farming systems, leading to improved livelihoods and economic stability.

In addition to boosting income, agroforestry systems contribute to **food security** by enhancing the availability, access, and utilization of diverse and nutritious food sources. Trees and shrubs integrated into agricultural landscapes can provide a variety of edible products, such as fruits, nuts, and leaves, which can supplement and diversify the diets of local communities. Furthermore, agroforestry practices can improve soil fertility and water availability, thereby increasing the productivity of adjacent food crops. In the face of climate change-induced disruptions to food systems, agroforestry's role in bolstering food security becomes all the more crucial.

Agroforestry also plays a significant role in fostering social cohesion and community resilience. By promoting the sustainable use of natural resources, agroforestry can help reduce conflicts over land and water, which are often exacerbated by climate change. Additionally, the implementation of agroforestry projects often involves collaboration among various stakeholders, including farmers, government agencies, non-governmental organizations, and research institutions. This collaborative approach fosters a sense of community ownership and shared responsibility, **strengthening social bonds** and enhancing the collective capacity to adapt to climate change.

Another significant socioeconomic benefit of agroforestry is its potential to create **employment opportunities**, particularly in rural areas. The labor-intensive nature of agroforestry practices, such as planting, pruning, and harvesting, generates jobs for local communities, thereby contributing to poverty alleviation and rural development. Moreover, the processing and marketing of agroforestry products can stimulate the growth of local industries and value chains, further enhancing economic opportunities and regional development.

In conclusion, the socioeconomic benefits of agroforestry in a changing climate are manifold and far-reaching. By diversifying income sources, enhancing food security, fostering social cohesion, and creating employment opportunities, agroforestry systems can significantly contribute to the well-being and resilience of communities in the face of climate change. As we strive to build a more sustainable and equitable future, embracing agroforestry as a holistic land management practice becomes a viable option and a necessary step toward achieving our collective goals.

Embracing Agroforestry as a Pathway to a Sustainable Future

In conclusion, it is evident that agroforestry and silvopasture play a significant role in mitigating the impacts of climate change while providing numerous environmental, social, and economic benefits. Therefore, as our world grapples with the consequences of a rapidly changing climate, we must explore and embrace sustainable land management practices such as agroforestry to ensure a resilient and thriving future for our planet and its inhabitants.

Agroforestry systems have demonstrated their potential to sequester carbon, enhance biodiversity, and promote ecosystem resilience. By integrating trees and shrubs with crops and livestock, these systems not only contribute to climate change mitigation but also provide a diverse array of ecosystem services that support human well-being and the **health of our planet.** Furthermore, agroforestry practices can help farmers adapt to the changing climate by improving soil fertility, reducing erosion, and providing shade and shelter for crops and animals.

The socioeconomic benefits of agroforestry are also noteworthy, as these systems can improve livelihoods, create job opportunities, and contribute to food security. Furthermore, by diversifying agricultural production and income sources, agroforestry can help farmers and rural communities become more resilient to the uncertainties of climate change and market fluctuations. Additionally, agroforestry can promote social equity and gender equality by empowering women and marginalized groups to participate in decision-making processes and benefit from the resources and opportunities provided by these systems.

As we move forward, policymakers, researchers, and practitioners must work together to promote the widespread adoption of agroforestry and silvopasture systems. This can be achieved through the development of supportive policies, increased investment in research and development, and the provision of technical assistance and training to farmers. Moreover, fostering collaboration and knowledge exchange among stakeholders, including farmers, researchers, and policymakers, is crucial to ensure that agroforestry practices are tailored to local contexts and address the specific needs and challenges of different communities.

In the face of climate change, embracing agroforestry as a pathway to a sustainable future is not only a wise decision but a necessary one. By integrating trees into our agricultural landscapes, we can harness the power of nature to **mitigate climate change**, enhance ecosystem resilience, and improve the well-being of people and the planet. As we continue to confront the challenges posed by a changing climate, let us remember that the solutions we seek may already be rooted in the ground beneath our feet.

Chapter Summary

- Agroforestry, a land management practice that integrates trees and shrubs with crops and livestock systems, plays a crucial role in climate change mitigation and adaptation.
- Agroforestry systems have the potential to sequester significant amounts of carbon, reducing greenhouse gas emissions and contributing to climate change mitigation.
- By enhancing biodiversity and ecosystem resilience, agroforestry systems create more diverse and adaptable ecosystems that can better withstand the impacts of climate change.

- Sustainable land management practices, such as conservation tillage, cover cropping, and agroforestry, can improve soil health, reduce erosion, and increase water infiltration, contributing to climate change adaptation.
- Agroforestry systems provide socioeconomic benefits, such as diversified income sources, improved livelihoods, and enhanced food security, helping communities adapt to climate change.
- The implementation of agroforestry practices can foster social cohesion and community resilience by promoting the sustainable use of natural resources and encouraging collaboration among stakeholders.
- Policymakers, researchers, and practitioners must work together to promote the widespread adoption of agroforestry systems through supportive policies, increased investment in research and development, and technical assistance and training for farmers.
- Embracing agroforestry as a pathway to a sustainable future is essential in addressing the challenges posed by climate change, as it harnesses the power of nature to mitigate climate change, enhance ecosystem resilience, and improve the well-being of people and the planet.

6

BIODIVERSITY CONSERVATION AND ECOSYSTEM SERVICES IN AGROFORESTRY

An image that showcases a thriving agroforestry landscape, complete with diverse plant and animal species coexisting harmoniously.

In today's rapidly changing world, the need for sustainable agricultural practices that promote biodiversity conservation and provide essential ecosystem services has never been more critical. Agroforestry, the intentional integration of trees and shrubs into crop and animal farming systems, offers a promising solution to this challenge. By combining the best of both worlds - agriculture and forestry - agroforestry systems can enhance biodiversity, improve soil health, and contribute to climate change mitigation, among other benefits.

This chapter delves into the vital role of agroforestry and silvopasture in **biodiversity conservation** and the **provision of ecosystem services.** We will begin by exploring the importance of biodiversity in agroforestry systems and how these systems can contribute to the conservation of various plant and animal species. Next, we will discuss the numerous ecosystem services that agroforestry provides, including carbon sequestration, soil fertility enhancement, and water regulation.

Silvopasture, a specific type of agroforestry that combines livestock grazing with the cultivation of trees, will be examined as a sustainable approach to biodiversity conservation. Finally, we will conclude by discussing the future of biodiversity conservation and ecosystem services in agroforestry, highlighting the potential of these systems to address some of the most pressing environmental challenges of our time.

As you read through this chapter, we invite you to consider the immense potential of agroforestry and silvopasture as tools for biodiversity conservation and ecosystem services provision. By understanding the benefits and challenges of these practices, we can work together to create a more sustainable, resilient, and biodiverse world for future generations.

The Role of Agroforestry in Enhancing Biodiversity

Agroforestry, the intentional integration of trees and shrubs into crop and animal farming systems, plays a vital role in enhancing biodiversity. By creating a more diverse and complex landscape, agroforestry systems promote a wide range of flora and fauna, contributing to the overall health and resilience of ecosystems. In this section, we will explore the various ways in which agroforestry contributes to biodiversity conservation and the benefits that arise from this symbiotic relationship.

One of the most significant ways in which agroforestry enhances biodiversity is by **providing habitat for various species.** Trees and shrubs offer shelter, nesting sites, and food sources for numerous birds, mammals, insects, and other organisms. This, in turn, creates a more balanced ecosystem, as the presence of diverse species helps to regulate pests and diseases, reducing the need for chemical interventions. Moreover, the integration of trees and shrubs within agricultural landscapes can serve as biological corridors, connecting fragmented habitats and allowing for the movement of species across the landscape.

Agroforestry systems also contribute to the conservation of genetic diversity, which is crucial for the long-term adaptability and survival of species. By cultivating various crops and tree species, agroforestry practices help preserve the genetic resources of both cultivated and wild plants. This is particularly important in the face of climate change, as genetic diversity enables species to adapt to changing environmental conditions and withstand various stressors.

In addition to conserving biodiversity, agroforestry systems also promote the **abundance and diversity of pollinators**, such as bees, butterflies, and other insects. Pollinators are essential for the reproduction of many plant species, including those grown for food, and their decline has been a growing concern worldwide. By providing diverse floral resources and nesting sites, agroforestry systems support healthy pollinator populations, which in turn contribute to increased crop yields and overall ecosystem health.

Furthermore, agroforestry practices can help to maintain and enhance soil biodiversity. The presence of trees and shrubs in agricultural systems promotes the growth of diverse microorganisms, such as bacteria and fungi, which play a crucial role in nutrient cycling and soil fertility. These microorganisms, along with other soil-dwelling organisms like earthworms, contribute to the breakdown of organic matter and the release of essential nutrients, ultimately supporting the growth of healthy crops.

In conclusion, agroforestry plays a critical role in enhancing biodiversity by providing habitat for various species, conserving genetic diversity, supporting pollinator populations, and maintaining soil biodiversity. The benefits of this increased biodiversity are numerous; they include improved ecosystem health, pest and disease regulation, and increased crop yields. Farmers and land managers can contribute to biodiversity conservation by adopting agroforestry practices while reaping the rewards of a more resilient and productive agricultural system.

Ecosystem Services Provided by Agroforestry Systems

Agroforestry systems, which integrate trees, shrubs, and crops, provide a wide range of ecosystem services that benefit both humans and the environment. These services are the direct and indirect contributions of ecosystems to human well-being, and they play a crucial role in sustaining the health and productivity of the planet. In this section, we will explore the various ecosystem services offered by agroforestry systems, emphasizing their importance in maintaining a balanced and thriving ecosystem.

First and foremost, agroforestry systems contribute to climate regulation by **sequestering carbon dioxide** from the atmosphere. Trees and other vegetation in these systems absorb and store carbon in their biomass, helping to mitigate the effects of climate change. Moreover, agroforestry practices can reduce greenhouse gas emissions from agriculture by decreasing the need for synthetic fertilizers and other chemical inputs.

Another vital ecosystem service that agroforestry systems provide is the **enhancement of soil fertility and quality**. Trees and shrubs in agroforestry systems improve soil structure by reducing erosion, increasing organic matter content, and promoting nutrient cycling. The deep root systems of trees also help to access nutrients from lower soil layers, making them available to nearby crops. Additionally, some tree species, such as legumes, can fix atmospheric nitrogen, further enriching the soil and reducing the need for synthetic fertilizers.

Agroforestry systems also contribute to **water regulation and purification.** The presence of trees and shrubs helps to reduce surface runoff, increase water infiltration, and recharge groundwater resources. This, in turn, prevents soil erosion, reduces the risk of flooding, and maintains water quality by filtering pollutants and sediment. Furthermore, agroforestry practices can enhance water use efficiency by providing shade and reducing evapotranspiration, which is particularly important in arid and semi-arid regions.

Biodiversity conservation is another significant ecosystem service offered by agroforestry systems. By creating a diverse and complex habitat, agroforestry practices support flora and fauna, pollinators, natural predators, and other beneficial organisms. This increased biodiversity can improve crop pollination, enhance pest control, and overall greater agroecosystem resilience.

Agroforestry systems also provide various provisioning services, such as food, fuel, and timber. By integrating trees and shrubs with crops, these systems can produce diverse products, including fruits, nuts, vegetables, and fodder for livestock. This diversification can improve food security and nutrition and provide additional income sources for farmers. Moreover, agroforestry practices can supply sustainable sources of fuelwood and timber, reducing pressure on natural forests and contributing to their conservation.

In conclusion, agroforestry systems offer many ecosystem services essential for the well-being of humans and the environment. By integrating trees, shrubs, and crops, these systems promote climate regulation, soil fertility, water management, biodiversity conservation, and the provision of various goods. As such, agroforestry practices represent a sustainable and resilient approach to land management, with the potential to address many of the challenges faced by modern agriculture.

Silvopasture: A Sustainable Approach to Biodiversity Conservation

Silvopasture, a unique and innovative agroforestry practice, combines the cultivation of trees, forage, and livestock in a single, integrated system. This sustainable approach optimizes land use and plays a vital role in biodiversity conservation. In this section, we will explore the various ways in which silvopasture contributes to the preservation of biodiversity and the overall health of ecosystems.

At its core, silvopasture is designed to **mimic natural ecosystems,** where multiple species coexist and interact harmoniously. By incorporating a diverse range of tree species, forage plants, and livestock breeds, silvopasture systems promote a rich and varied habitat that supports a wide array of flora and fauna. This diversity is essential for maintaining the ecological balance and resilience of the system, as it reduces the risk of disease outbreaks, pest infestations, and other threats to biodiversity.

One of the most significant benefits of silvopasture is its ability to **provide habitat for various wildlife species.** The integration of trees and shrubs within grazing lands creates a more complex and heterogeneous landscape, offering shelter, food, and nesting sites for birds, mammals, and insects. This, in turn, increases the overall species richness and abundance within the system, contributing to the preservation of local and regional biodiversity.

In addition to providing habitat for wildlife, silvopasture systems also contribute to biodiversity conservation by **enhancing the connectivity of fragmented landscapes.** As agricultural expansion and urbanization continue to encroach upon natural habitats, the creation of silvopasture systems can serve as vital corridors for wildlife movement between isolated patches of forest or grassland. This connectivity is crucial for the long-term survival of many species, as it allows for the exchange of genetic material and the recolonization of vacant habitats.

Moreover, silvopasture systems are crucial in **maintaining and enhancing soil biodiversity.** The presence of trees and shrubs within grazing lands helps to prevent soil erosion, improve soil structure, and increase the availability of nutrients. These factors, in turn, support a diverse and active community of soil organisms, including bacteria, fungi, and invertebrates. These organisms play a critical role in nutrient cycling, decomposition, and other essential ecosystem processes, ultimately contributing to the overall health and productivity of the system.

In conclusion, silvopasture represents a sustainable and promising approach to biodiversity conservation within agricultural landscapes. By integrating trees, forage, and livestock in a single system, silvopasture not only optimizes land use but also promotes a rich and varied habitat that supports a wide array of species. Furthermore, as the global demand for food continues to grow, adopting silvopasture practices can help ensure the long-term health and resilience of our ecosystems while also providing valuable ecosystem services to farmers and society as a whole.

The Future of Biodiversity Conservation and Ecosystem Services in Agroforestry

In conclusion, the future of biodiversity conservation and ecosystem services in agroforestry is promising and full of potential. As the world grapples with the challenges of climate change, food security, and environmental degradation, agroforestry and silvopasture practices offer sustainable solutions that can help address these pressing issues. By integrating trees, crops, and livestock, these systems can enhance biodiversity, improve soil health, and provide a range of ecosystem services that benefit both people and the environment.

The role of agroforestry in enhancing biodiversity cannot be overstated. By providing diverse habitats and resources for a wide array of species, agroforestry systems can support the conservation of flora and fauna. This, in turn, contributes to the overall health and resilience of ecosystems, which is crucial in the face of ongoing environmental challenges.

Ecosystem services provided by agroforestry systems are also of immense value. These services include carbon sequestration, water regulation, erosion control, and pollination, among others. By harnessing these benefits, agroforestry can help mitigate climate change, protect water resources, and support agricultural productivity. Furthermore, these services have the potential to generate economic value for farmers and landowners, providing incentives for the adoption of sustainable practices.

Silvopasture, as a specific form of agroforestry, offers a sustainable approach to biodiversity conservation by integrating livestock with trees and pasture. This practice can help maintain and restore degraded landscapes, reduce pressure on natural habitats, and support the conservation of species that depend on these ecosystems. Moreover, silvopasture can contribute to the livelihoods of farmers and rural communities by diversifying income sources and enhancing the resilience of agricultural systems.

The case studies presented in this chapter demonstrate the effectiveness of agroforestry and silvopasture practices in promoting biodiversity conservation and delivering ecosystem services. These examples inspire and guide the development and implementation of similar practices in different contexts around the world.

Looking ahead, the future of biodiversity conservation and ecosystem services in agroforestry depends on the continued research, innovation, and dissemination of knowledge in this field. Policymakers, researchers, and practitioners must work together to develop supportive policies, incentives, and educational programs that encourage the widespread adoption of agroforestry and silvopasture practices. By doing so, we can ensure that these systems play a central role in addressing the global challenges of our time and contribute to a more sustainable and resilient future for our planet.

Chapter Summary

- Agroforestry, the intentional integration of trees and shrubs into crop and animal farming systems, offers a sustainable solution to promote biodiversity conservation and provide essential ecosystem services.

- Agroforestry enhances biodiversity by providing habitat for various species, conserving genetic diversity, supporting pollinator populations, and maintaining soil biodiversity.
- Ecosystem services provided by agroforestry systems include climate regulation through carbon sequestration, soil fertility enhancement, water regulation and purification, biodiversity conservation, and the provision of various goods such as food, fuel, and timber.
- Silvopasture, a specific type of agroforestry that combines livestock grazing with the cultivation of trees, is a sustainable approach to biodiversity conservation that mimics natural ecosystems and promotes a rich and varied habitat.
- The future of biodiversity conservation and ecosystem services in agroforestry is promising, with the potential to address pressing global challenges such as climate change, food security, and environmental degradation.
- Continued research, innovation, and dissemination of knowledge in agroforestry and silvopasture are essential for their widespread adoption and success in promoting biodiversity conservation and delivering ecosystem services.
- Policymakers, researchers, and practitioners must collaborate to develop supportive policies, incentives, and educational programs that encourage the adoption of agroforestry and silvopasture practices, contributing to a more sustainable and resilient future for our planet.

7

SOCIOECONOMIC BENEFITS OF AGROFORESTRY AND SILVOPASTURE

A futuristic city skyline with agroforestry and silvopasture systems integrated into the urban environment.

In recent years, the world has witnessed a growing interest in sustainable agricultural practices that not only ensure food security but also contribute to the socioeconomic development of rural communities. Among these practices, agroforestry and silvopasture have emerged as promising solutions to address the challenges of climate change, land degradation, and rural poverty. By integrating trees, crops, and livestock on the same land, these systems offer a multitude of benefits that can significantly improve the livelihoods of millions of people, particularly in developing countries.

Agroforestry, which combines the cultivation of trees with crops, and silvopasture, which integrates trees with livestock grazing, have been practiced for centuries in various parts of the world. However, their potential to contribute to socioeconomic development has only recently been recognized and appreciated. This chapter aims to explore the various ways in which agroforestry and silvopasture can enhance the well-being of individuals, communities, and nations by fostering income diversification, poverty reduction, food security, employment opportunities, and environmental services.

The following sections will delve into the specific socioeconomic benefits of agroforestry and silvopasture practices, highlighting their potential to transform the lives of farmers and rural communities. We will also discuss the role of these systems in addressing global challenges such as climate change and land degradation and how they can contribute to the achievement of the United Nations Sustainable Development Goals (SDGs).

By the end of this chapter, readers will gain a deeper understanding of the growing importance of agroforestry and silvopasture in socioeconomic development and how these practices can be harnessed to create a more sustainable, equitable, and prosperous world for all.

Income Diversification and Poverty Reduction

In recent years, agroforestry and silvopasture practices have emerged as viable and sustainable alternatives to conventional agriculture, offering significant socioeconomic benefits to rural communities. One of the most notable advantages of these integrated land management systems is the potential for income diversification and poverty reduction. By combining the cultivation of trees, crops, and livestock, farmers can tap into **multiple revenue streams**, thereby reducing their vulnerability to market fluctuations and environmental shocks.

Agroforestry and silvopasture practices offer a myriad of income-generating opportunities for farmers. For instance, the integration of trees in agricultural landscapes can provide timber, fuelwood, and non-timber forest products (NTFPs) such as fruits, nuts, and medicinal plants. These products can be sold in local and international markets, generating additional income for farmers and contributing to the overall economic growth of rural areas.

Moreover, integrating livestock in agroforestry and silvopasture systems can lead to **increased productivity and profitability.** Livestock can graze on the understory vegetation, reducing the need for external feed inputs and lowering production costs. At the same time, the presence of animals can improve soil fertility through the deposition of manure, which in turn can enhance crop yields. This symbiotic relationship between trees, crops, and livestock creates a more resilient and diversified farming system that can better withstand economic and environmental challenges.

Another important aspect of income diversification through agroforestry and silvopasture practices is the potential for value addition and niche market development. By adopting these sustainable land management systems, farmers can differentiate their products in the market, catering to the growing consumer demand for environmentally friendly and socially responsible goods. This can lead to premium prices and increased profitability for farmers who embrace agroforestry and silvopasture practices.

Furthermore, agroforestry and silvopasture systems can contribute to poverty reduction by enhancing the resilience of rural communities to climate change and other environmental risks. By promoting biodiversity and ecosystem services, these practices can help buffer against the impacts of extreme weather events, pests, and diseases. This increased resilience can translate into more stable and secure livelihoods for farmers and their families, reducing their vulnerability to poverty.

In conclusion, agroforestry and silvopasture practices hold great promise for income diversification and poverty reduction in rural areas. By integrating trees, crops, and livestock, farmers can tap into multiple revenue streams, enhance their resilience to economic and environmental shocks, and capitalize on niche market opportunities. As the world grapples with the challenges of climate change, food insecurity, and rural poverty, the adoption of agroforestry and silvopasture practices can play a crucial role in fostering sustainable and inclusive socioeconomic development.

Enhancing Food Security and Nutrition through Integrated Land Management

The integration of agroforestry and silvopasture practices within land management systems has the potential to significantly enhance food security and nutrition for communities worldwide. By combining the cultivation of trees, crops, and livestock, these systems offer a diverse range of agricultural products that can contribute to a balanced and nutritious diet. Furthermore, integrated land management practices can help to increase the resilience of agricultural systems, ensuring a more stable and reliable food supply in the face of environmental challenges and climate change.

One of the primary ways in which agroforestry and silvopasture practices contribute to food security is through the **diversification of agricultural production**. By incorporating a variety of tree species, crops, and livestock into a single system, farmers can reduce their reliance on a single commodity and mitigate the risks associated with crop failure or market fluctuations.

This diversification provides a more stable income for farmers and ensures a more diverse and nutritious diet for local communities. For example, the integration of fruit and nut trees into agricultural systems can provide essential vitamins and minerals, while the incorporation of leguminous trees can improve soil fertility and support the production of protein-rich crops such as beans and peas.

In addition to diversifying agricultural production, agroforestry and silvopasture practices can also contribute to enhanced food security by increasing the overall productivity of the land. By optimizing the use of available resources such as sunlight, water, and nutrients, these integrated systems can yield higher outputs than traditional monoculture systems. For instance, the strategic placement of trees within a crop field can create microclimates that improve growing conditions for both crops and livestock, leading to increased yields and improved animal health. Moreover, integrating trees and livestock can facilitate nutrient cycling, reducing the need for external inputs such as chemical fertilizers and supporting more sustainable and productive agricultural systems.

Employing agroforestry and silvopasture practices can also help to address the challenges posed by climate change, which threatens global food security. These integrated systems can enhance the resilience of agricultural landscapes by providing ecosystem services such as erosion control, water regulation, and carbon sequestration. By promoting biodiversity and improving soil health, agroforestry and silvopasture systems can help to buffer the impacts of extreme weather events and support the long-term productivity of agricultural land.

In conclusion, the adoption of agroforestry and silvopasture practices within integrated land management systems has the potential to significantly enhance food security and nutrition for communities around the world. By diversifying agricultural production, increasing land productivity, and promoting resilience in the face of climate change, these systems offer a sustainable and effective solution to some of the most pressing challenges facing global food systems today. As we continue to explore the full potential of agroforestry and silvopasture for socioeconomic benefits, we must prioritize the integration of these practices into land management strategies and support their widespread adoption by farmers and land managers worldwide.

Employment Opportunities and Rural Development in Agroforestry and Silvopasture Systems

The potential of agroforestry and silvopasture systems to create employment opportunities and foster rural development cannot be overstated. As these integrated land management practices gain traction, they offer myriad possibilities for job creation, skill development, and community empowerment, particularly in rural areas where traditional agricultural practices may have reached their limits.

One of the most significant aspects of agroforestry and silvopasture systems is their ability to **generate diverse income streams.** By integrating trees, crops, and livestock, farmers can tap into multiple markets, thereby reducing their vulnerability to fluctuations in commodity prices and seasonal variations. This diversification not only increases the overall profitability of farming operations but also creates a wide range of job opportunities for individuals with different skill sets. For instance, the management of tree nurseries, crop cultivation, and animal husbandry all require specialized knowledge and expertise, which in turn creates demand for skilled labor.

Moreover, agroforestry and silvopasture systems often involve the cultivation of **high-value tree species**, such as fruit and nut trees, timber, and medicinal plants. These products can fetch premium prices in local and international markets, providing farmers with additional sources of income and creating opportunities for value-added processing and marketing. This can lead to the establishment of small and medium-sized enterprises (SMEs) in rural areas, which in turn generate employment and stimulate local economies.

Another essential aspect of agroforestry and silvopasture systems is their potential to **enhance rural development** by promoting social cohesion and community empowerment. As these practices often require cooperation and coordination among farmers, they can foster a sense of collective responsibility and shared ownership of natural resources. This can lead to the formation of farmer cooperatives, associations, and other community-based organizations that can help farmers access credit, training, and other resources necessary to successfully implement agroforestry and silvopasture practices.

Furthermore, agroforestry and silvopasture systems can contribute to rural development by improving the overall quality of life in rural areas. By enhancing the productivity and sustainability of agricultural landscapes, these practices can ensure a stable supply of food, fuel, and other essential goods and services for local communities. Additionally, the integration of trees and livestock into farming systems can provide significant ecological benefits, such as improved soil fertility, water quality, and biodiversity, which in turn can enhance the aesthetic and recreational value of rural landscapes.

In conclusion, the adoption of agroforestry and silvopasture systems can play a crucial role in creating employment opportunities and fostering rural development. By diversifying income streams, promoting skill development, and encouraging community cooperation, these integrated land management practices can help to revitalize rural economies and improve the overall quality of life for millions of people around the world. As we grapple with the challenges of climate change, population growth, and resource scarcity, we must recognize and harness the full potential of agroforestry and silvopasture for socioeconomic benefits.

Environmental Services and Climate Change Mitigation: The Role of Agroforestry and Silvopasture

As the world grapples with the mounting challenges of climate change, the importance of sustainable land management practices, such as agroforestry and silvopasture, cannot be overstated. These integrated systems contribute to socioeconomic development and provide vital environmental services that help mitigate the impacts of climate change. In this section, we will explore the various ways in which agroforestry and silvopasture contribute to environmental conservation and climate change mitigation.

One of the most significant environmental services provided by agroforestry and silvopasture systems is **carbon sequestration.** Trees and other perennial vegetation in these systems absorb and store atmospheric carbon dioxide, effectively reducing greenhouse gas emissions. This process is particularly important in the context of climate change, as increased levels of carbon dioxide in the atmosphere contribute to global warming. By incorporating trees and shrubs into agricultural landscapes, agroforestry and silvopasture practices can help offset carbon emissions and combat climate change.

In addition to carbon sequestration, agroforestry and silvopasture systems contribute to **climate change mitigation** by promoting biodiversity and ecosystem resilience. The integration of trees, crops, and livestock creates diverse habitats that support a wide range of plant and animal species. This biodiversity is crucial for maintaining ecosystem stability and resilience in the face of climate change, as it enables ecosystems to better adapt to and recover from disturbances such as droughts, floods, and pest outbreaks.

Furthermore, agroforestry and silvopasture systems can help reduce soil erosion and improve water quality. Tree roots stabilize soils, while their foliage intercepts rainfall, reducing the impact of heavy rains on the soil surface. This, in turn, minimizes soil erosion and sedimentation in water bodies, protecting aquatic ecosystems and maintaining water quality. Moreover, the presence of trees and other vegetation in these systems can enhance water infiltration and groundwater recharge, contributing to more sustainable water management in the face of climate change-induced water scarcity.

Another important environmental service provided by agroforestry and silvopasture systems is the **regulation of microclimates**. Trees and other vegetation can moderate temperature extremes and create more favorable conditions for crop and livestock production. For instance, tree shade can reduce heat stress on livestock, improving their productivity and welfare. Similarly, the presence of trees can buffer crops from extreme temperatures, reducing the risk of crop failure due to heat or frost.

In conclusion, agroforestry and silvopasture systems offer a multitude of environmental services that contribute to climate change mitigation and adaptation. By promoting carbon sequestration, biodiversity, soil conservation, water quality, and microclimate regulation, these integrated land management practices can play a crucial role in building a more sustainable and resilient future for our planet. As we continue to face the challenges of climate change, we must recognize and harness the full potential of agroforestry and silvopasture for both socioeconomic development and environmental conservation.

Realizing the Full Potential of Agroforestry and Silvopasture for Socioeconomic Benefits

In conclusion, agroforestry and silvopasture have emerged as powerful tools for socioeconomic development, offering many benefits beyond traditional agricultural practices. By integrating trees, crops, and livestock, these systems not only support income diversification and poverty reduction but also contribute to food security, nutrition, employment opportunities, and rural development.

To fully realize the potential of agroforestry and silvopasture, it is crucial to **promote their adoption and implementation on a larger scale.** This can be achieved through various means, such as raising awareness among farmers and landowners about the benefits of these practices, providing financial incentives and technical assistance, and encouraging research and innovation in the field.

Moreover, it is essential to recognize the role of agroforestry and silvopasture in addressing environmental challenges, particularly climate change. By sequestering carbon, conserving biodiversity, and enhancing ecosystem services, these systems can significantly contribute to global efforts to mitigate and adapt to the impacts of climate change. Therefore, policymakers and stakeholders must prioritize the integration of agroforestry and silvopasture into national and international climate change strategies.

In addition, fostering collaboration among various sectors, such as agriculture, forestry, and livestock, is key to maximizing the socioeconomic benefits of agroforestry and silvopasture. This can be facilitated by establishing cross-sectoral platforms, networks, and partnerships that promote knowledge exchange, capacity building, and joint action.

Finally, it is important to acknowledge that the success of agroforestry and silvopasture depends on the active participation and empowerment of local communities. By involving farmers, landowners, and other stakeholders in the planning, implementation, and monitoring of these systems, we can ensure that their needs and preferences are taken into account and that the benefits of agroforestry and silvopasture are equitably distributed.

In summary, agroforestry and silvopasture have the potential to transform the way we manage our land and resources, leading to more sustainable, resilient, and inclusive rural landscapes. By embracing these practices and working together, we can create a brighter future for our planet and its people.

Chapter Summary

- Agroforestry and silvopasture are sustainable agricultural practices that integrate trees, crops, and livestock, offering numerous socioeconomic benefits and addressing challenges such as climate change, land degradation, and rural poverty.
- These practices contribute to income diversification and poverty reduction by providing multiple revenue streams, reducing vulnerability to market fluctuations and environmental shocks, and creating opportunities for value addition and niche market development.
- Agroforestry and silvopasture enhance food security and nutrition by diversifying agricultural production, increasing land productivity, and promoting resilience in the face of climate change.
- These integrated land management systems create employment opportunities and foster rural development by generating diverse income streams, promoting skill development, and encouraging community cooperation.

- Agroforestry and silvopasture contribute to climate change mitigation by providing environmental services such as carbon sequestration, biodiversity conservation, soil erosion control, water quality improvement, and microclimate regulation.
- To fully realize the potential of these practices, it is crucial to promote their adoption and implementation on a larger scale through awareness-raising, financial incentives, technical assistance, and research and innovation.
- Integrating agroforestry and silvopasture into national and international climate change strategies is essential, as is fostering collaboration among various sectors such as agriculture, forestry, and livestock.
- The success of agroforestry and silvopasture depends on the active participation and empowerment of local communities, ensuring that their needs and preferences are taken into account and that the benefits are equitably distributed.

8

CHALLENGES AND LIMITATIONS OF AGROFORESTRY AND SILVOPASTURE IMPLEMENTATION

A futuristic scenario where agroforestry and silvopasture systems have overcome their implementation challenges and have become the norm in sustainable agriculture.

In recent years, the world has witnessed an increasing awareness of the importance of sustainable agricultural practices. Amidst the challenges of climate change, population growth, and food security, agroforestry and silvopasture have emerged as promising solutions to address these pressing issues. These innovative land management systems integrate trees, crops, and livestock, creating a harmonious and productive ecosystem that benefits both the environment and the people who depend on it. As we delve into the challenges and limitations of implementing agroforestry and silvopasture, we must understand their growing significance in the quest for a sustainable future.

Agroforestry, the intentional integration of trees and shrubs into crop and animal farming systems, offers a myriad of environmental, social, and economic benefits. By incorporating trees into agricultural landscapes, agroforestry practices help to enhance soil fertility, reduce soil erosion, and improve water quality. Additionally, these systems provide habitat for diverse flora and fauna, contributing to increased biodiversity and ecosystem resilience. From a social perspective, agroforestry can improve livelihoods by providing farmers with diversified income sources, such as timber, fruits, and medicinal plants. Furthermore, the integration of trees in farming systems can contribute to climate change mitigation by sequestering carbon dioxide from the atmosphere.

Silvopasture, a subset of agroforestry, involves the integration of trees, forage, and livestock in a single management system. This approach offers numerous benefits for both the environment and the farmer. By providing shade and shelter, trees can improve animal welfare and productivity while enhancing the landscape's overall aesthetic appeal. The presence of livestock can help control weeds and pests, reducing the need for chemical inputs. Moreover, silvopasture systems can increase the overall productivity of the land, providing additional income streams for farmers through the sale of timber, fruits, and other tree products.

Despite the numerous advantages of agroforestry and silvopasture, their widespread adoption faces several challenges and limitations. These obstacles range from land tenure and property rights issues to financial and technical constraints, as well as environmental and climatic factors and socio-cultural barriers. In the following sections, we will explore these challenges in detail and discuss potential strategies to overcome them, paving the way for a more sustainable and resilient agricultural future.

Land Tenure and Property Rights: Obstacles to Adoption

As the world increasingly recognizes the importance of agroforestry and silvopasture in promoting sustainable agriculture, land tenure and property rights have emerged as significant obstacles to their widespread adoption. In this section, we delve into the complexities of land ownership and usage rights and how they can hinder the integration of agroforestry and silvopasture practices.

Land tenure, the way people access and use land, plays a crucial role in determining the success of agroforestry and silvopasture initiatives. Secure land tenure gives farmers the confidence to invest in long-term, sustainable land management practices, such as planting trees and integrating livestock into their farming systems. However, in many parts of the world, land tenure systems are characterized by insecurity, inequality, and a lack of clear property rights.

One of the main challenges associated with land tenure is the **prevalence of customary or informal land tenure systems**, particularly in developing countries. Under these systems, land rights are often based on local traditions and customs rather than formal legal frameworks. This can result in a lack of clarity regarding land ownership and usage rights, making it difficult for farmers to invest in agroforestry and silvopasture practices. Additionally, customary land tenure systems can perpetuate gender inequality, as women may have limited access to land and decision-making power, further complicating the adoption of sustainable land management practices.

Another obstacle to the adoption of agroforestry and silvopasture is the **fragmentation of land ownership.** Small landholdings can limit the potential for implementing these practices, as farmers may not have sufficient land to establish tree-based systems or integrate livestock effectively. Moreover, fragmented land ownership can hinder collaboration among farmers, which is essential for the successful implementation of agroforestry and silvopasture practices at a landscape scale.

In some cases, government policies and regulations can impede the adoption of agroforestry and silvopasture. For instance, land-use policies that prioritize monoculture crop production or restrict tree planting on agricultural land can discourage farmers from adopting these practices. Furthermore, unclear or restrictive regulations regarding the use of common resources, such as forests and grazing lands, can limit the potential for integrating trees and livestock into farming systems.

In order to overcome these obstacles, it is essential to promote secure land tenure and clear property rights that encourage investment in sustainable land management practices. This can be achieved through legal and policy reforms that recognize and formalize customary land rights, promote gender equality in land ownership, and facilitate collaboration among landowners. Additionally, governments and development organizations should prioritize capacity-building efforts to help farmers navigate complex land tenure systems and advocate for their rights.

In conclusion, land tenure and property rights are critical factors influencing the adoption of agroforestry and silvopasture practices. Addressing these challenges is essential for unlocking the full potential of these sustainable land management practices, ultimately contributing to a more resilient and productive agricultural landscape.

Financial and Technical Constraints: The Struggle for Resources

The implementation of agroforestry and silvopasture practices is not without its challenges, and among the most significant of these are the financial and technical constraints faced by farmers and land managers. In this section, we will explore the various financial and technical barriers that hinder the widespread adoption of these sustainable land management practices and discuss potential solutions to overcome these obstacles.

One of the primary financial constraints faced by farmers interested in adopting agroforestry and silvopasture practices is the **initial investment** required to establish these systems. This includes the cost of purchasing and planting trees, installing fencing and other infrastructure, and acquiring the necessary equipment and tools. Additionally, there may be a significant time lag between the initial investment and the realization of economic returns from these practices, as trees take several years to mature and produce marketable products. This can be particularly challenging for small-scale farmers with limited financial resources, who may struggle to secure the necessary capital to invest in these systems.

Another financial challenge is the **lack of access to credit** and financing options for agroforestry and silvopasture projects. Traditional lending institutions may be unfamiliar with these practices and may perceive them as high-risk investments, making it difficult for farmers to secure loans or other forms of financial support. This can be further exacerbated by the absence of well-developed markets for agroforestry and silvopasture products, which can make it difficult for farmers to predict future income streams and plan their investments accordingly.

In addition to financial constraints, **technical barriers** can also impede the adoption of agroforestry and silvopasture practices. Many farmers and land managers may lack the necessary knowledge and skills to successfully implement and manage these systems, particularly in regions where these practices are not yet widespread. This can include a lack of understanding of the appropriate tree species to plant, the optimal planting densities and arrangements, and the best management practices to maximize both agricultural and forestry yields.

In order to overcome these financial and technical constraints, several strategies can be employed. First, **targeted training and extension programs** can help to build the capacity of farmers and land managers to successfully implement and manage agroforestry and silvopasture systems. This can include the development of context-specific training materials and the provision of technical assistance through extension agents or farmer-to-farmer networks.

Second, innovative **financing mechanisms** can be developed to support the adoption of agroforestry and silvopasture practices. This could include establishing dedicated loan programs or credit guarantees for agroforestry and silvopasture projects, as well as the development of novel financial instruments such as payment for ecosystem services schemes or carbon credits that can help to monetize the environmental benefits of these practices.

Lastly, efforts should be made to **strengthen and develop markets** for agroforestry and silvopasture products, both locally and internationally. This can include the promotion of certification schemes and labeling initiatives that recognize and reward sustainable land management practices, as well as the establishment of market linkages that connect producers with buyers and consumers.

In conclusion, while financial and technical constraints pose significant challenges to the widespread adoption of agroforestry and silvopasture practices, targeted interventions and innovative solutions can help to overcome these barriers and unlock the full potential of these sustainable land management systems. By addressing these constraints, we can move closer to a more sustainable and resilient future for our agricultural landscapes and the communities that depend on them.

Environmental and Climatic Factors: Adapting to Change

As the world grapples with the ever-increasing impacts of climate change, the importance of adopting sustainable agricultural practices such as agroforestry and silvopasture cannot be overstated. However, the implementation of these systems is not without its challenges, particularly when it comes to environmental and climatic factors. This section will delve into the various environmental and climatic obstacles that must be overcome to successfully integrate agroforestry and silvopasture into modern agriculture.

One of the most pressing environmental challenges facing agroforestry and silvopasture is the **need to adapt to changing climatic conditions.** As global temperatures continue to rise, the frequency and intensity of extreme weather events such as droughts, storms, and floods are expected to increase. These events can devastate agricultural systems, particularly those that rely on the delicate balance between trees, crops, and livestock. In order to ensure the long-term viability of agroforestry and silvopasture systems, farmers must be prepared to adapt their practices to accommodate these changing conditions.

One way to achieve this is by selecting tree and crop species that are well-suited to the specific environmental conditions of a given region. For example, in areas prone to drought, farmers may opt to plant drought-tolerant tree species such as acacia or mesquite, which can provide valuable shade and fodder for livestock while also helping to conserve soil moisture. Similarly, in regions with heavy rainfall, farmers might choose to plant fast-growing tree species with extensive root systems, such as willows or poplars, to help prevent soil erosion and improve water infiltration.

Another environmental challenge facing agroforestry and silvopasture systems is the need to maintain biodiversity and ecosystem health. While these systems have the potential to support a wide variety of plant and animal species, the introduction of non-native or invasive species can have detrimental effects on local ecosystems. To mitigate this risk, farmers should prioritize the use of **native tree and crop species** whenever possible and carefully monitor their land for signs of invasive species.

In addition to these environmental challenges, agroforestry and silvopasture systems must also contend with the issue of climate change mitigation. While these systems have been shown to be effective at sequestering carbon and reducing greenhouse gas emissions, there is still much work to be done in terms of quantifying and optimizing these benefits. For example, more research is needed to determine the optimal tree species, planting densities, and management practices for maximizing carbon sequestration in different agroforestry and silvopasture systems.

In conclusion, the successful implementation of agroforestry and silvopasture systems requires a deep understanding of the environmental and climatic factors at play and a willingness to adapt and innovate in the face of change. By selecting appropriate tree and crop species, prioritizing biodiversity and ecosystem health, and working to optimize climate change mitigation benefits, farmers can help ensure the long-term sustainability of these valuable agricultural systems.

Socio-Cultural Barriers: Overcoming Traditional Mindsets

As we delve deeper into the challenges of implementing agroforestry and silvopasture, it is crucial to recognize the socio-cultural barriers that often hinder the adoption of these sustainable practices. Traditional mindsets and cultural norms play a significant role in shaping the way communities interact with their environment and manage their resources. In this section, we will explore some of the socio-cultural barriers that impede the widespread adoption of agroforestry and silvopasture, and discuss potential strategies to overcome them.

One of the primary socio-cultural barriers to agroforestry and silvopasture adoption is the **resistance to change**. Many farming communities have been practicing conventional agriculture for generations, and the idea of integrating trees and livestock into their farming systems may seem unfamiliar and risky. This resistance to change can be attributed to a lack of knowledge about the benefits of agroforestry and silvopasture, as well as a fear of potential negative impacts on their livelihoods. In order to overcome this barrier, it is essential to **raise awareness** about the long-term economic, social, and environmental benefits of these practices through education, training, and demonstration projects.

Another socio-cultural barrier is the **gender dynamics within farming communities**. In many cases, women play a crucial role in managing natural resources, yet their voices and needs are often overlooked in decision-making processes. This can result in the neglect of agroforestry and silvopasture practices that could benefit both women and their communities. In order to address this issue, it is important to **promote gender equality** and empower women in the decision-making process, ensuring that their perspectives and needs are considered when planning and implementing agroforestry and silvopasture projects.

The third socio-cultural barrier is the **lack of cooperation and collaboration** among stakeholders within the community, such as farmers, landowners, and local authorities. Agroforestry and silvopasture practices often require collective action and shared responsibility to be successful. However, conflicting interests, power dynamics, and mistrust can hinder the establishment of effective partnerships. In order to overcome this barrier, it is vital to foster a sense of **community ownership** and shared responsibility for the success of agroforestry and silvopasture projects. This can be achieved through participatory planning, capacity building, and the establishment of local institutions that promote collaboration and collective decision-making.

Lastly, traditional beliefs and cultural practices can sometimes pose challenges to the adoption of agroforestry and silvopasture. For instance, certain tree species may have a cultural or religious significance, which may restrict their use in agroforestry systems. In such cases, it is crucial to respect and acknowledge these cultural values while promoting the benefits of agroforestry and silvopasture. This can be done by engaging local communities in selecting appropriate tree species and designing **culturally sensitive agroforestry and silvopasture** systems.

In conclusion, overcoming socio-cultural barriers is a critical step towards the successful implementation of agroforestry and silvopasture practices. By raising awareness, promoting gender equality, fostering collaboration, and respecting cultural values, we can pave the way for a more sustainable and resilient future for both people and the environment.

Overcoming Challenges for a Sustainable Future

As we have explored throughout this chapter, agroforestry and silvopasture systems hold immense potential for addressing a myriad of global challenges, including food security, climate change, and biodiversity loss. However, the successful implementation of these systems is not without its obstacles. Land tenure issues, financial and technical constraints, environmental and climatic factors, and socio-cultural barriers all pose significant challenges to the widespread adoption of agroforestry and silvopasture practices.

Despite these challenges, there is a reason for optimism. Innovative solutions and collaborative efforts are emerging to address these barriers, paving the way for a more sustainable and resilient future. To overcome land tenure and property rights issues, governments and organizations can work together to develop policies that support agroforestry and silvopasture initiatives while also ensuring equitable access to land and resources for smallholder farmers and indigenous communities.

Financial and technical constraints can be addressed through increased investment in research and development, as well as the establishment of support networks and extension services to provide farmers with the necessary knowledge and tools to implement these systems effectively. Public and private sector partnerships can play a crucial role in mobilizing resources and fostering innovation in this area.

Adapting to environmental and climatic factors requires a flexible and context-specific approach. By integrating traditional ecological knowledge with modern scientific research, agroforestry and silvopasture systems can be tailored to suit the unique conditions of a given region, ensuring their long-term viability and resilience in the face of a changing climate.

Finally, overcoming socio-cultural barriers necessitates **a shift in mindset** and a willingness to embrace change. Education and awareness-raising campaigns can help to challenge traditional notions of agriculture and promote the benefits of agroforestry and silvopasture systems. We can create a more inclusive and sustainable agricultural landscape by engaging local communities and fostering a sense of ownership and pride in these practices.

In conclusion, the challenges and limitations of agroforestry and silvopasture implementation are indeed significant, but they are not insurmountable. By working together to address these barriers, we can unlock the full potential of these systems and contribute to a more sustainable and resilient future for our planet and its inhabitants. The journey may be long and arduous, but the rewards are well worth the effort.

Chapter Summary

- Agroforestry and silvopasture are sustainable land management systems that integrate trees, crops, and livestock, offering environmental, social, and economic benefits.
- Land tenure and property rights issues, such as insecure land ownership and fragmented landholdings, can hinder the adoption of agroforestry and silvopasture practices.
- Financial and technical constraints, including the initial investment required and lack of access to credit, pose challenges for farmers interested in adopting these systems.
- Environmental and climatic factors, such as adapting to changing weather conditions and maintaining biodiversity, are crucial for the long-term viability of agroforestry and silvopasture systems.
- Socio-cultural barriers, such as resistance to change, gender dynamics, and lack of cooperation among stakeholders, can impede the widespread adoption of these practices.
- Overcoming these challenges requires innovative solutions, such as legal and policy reforms, targeted training and extension programs, and the development of novel financial instruments.

- Adapting to environmental and climatic factors requires a flexible and context-specific approach, integrating traditional ecological knowledge with modern scientific research.
- Addressing socio-cultural barriers necessitates a shift in mindset, education, and awareness-raising campaigns to promote the benefits of agroforestry and silvopasture systems and engage local communities.

9

CASE STUDIES: SUCCESSFUL AGROFORESTRY AND SILVOPASTURE SYSTEM

A futuristic scenario where agroforestry and silvopasture systems have overcome their implementation challenges and have become the norm in sustainable agriculture.

In recent years, the importance of sustainable agricultural practices has gained significant attention as the world grapples with the challenges of climate change, food security, and environmental degradation. Agroforestry and silvopasture are two such practices that have emerged as promising solutions to these pressing issues. By integrating trees and shrubs into agricultural landscapes, these systems offer a myriad of benefits, including increased biodiversity, improved soil health, and enhanced carbon sequestration. This chapter delves into the fascinating world of agroforestry and silvopasture by exploring various case studies that showcase the successful implementation of these systems around the globe.

The purpose of this chapter is to provide readers with a comprehensive understanding of the diverse agroforestry and silvopasture systems that have been employed in different parts of the world. By examining these case studies, we hope to inspire farmers, landowners, and policymakers to adopt these sustainable practices and contribute to the global effort toward a greener and more resilient future.

Join us as we embark on this exciting exploration of agroforestry and silvopasture case studies, and discover the transformative potential of these sustainable agricultural practices.

Innovative Agroforestry Systems Around the World

Agroforestry, the practice of integrating trees and shrubs into agricultural systems, has gained recognition as a sustainable and innovative approach to land management. This section will explore some of the most groundbreaking and successful agroforestry systems from around the world, showcasing the diversity and adaptability of these practices.

The Homegarden System in Kerala, India

In the southern Indian state of Kerala, the ancient practice of homegardens has been passed down through generations. These biodiverse, multi-layered agroforestry systems consist of a variety of trees, shrubs, and crops, providing food, medicine, and timber for the local communities. The homegardens of Kerala are a shining example of how agroforestry can promote food security, biodiversity, and climate resilience.

The Quesungual System in Honduras

The Quesungual system, developed by small-scale farmers in western Honduras, is a unique agroforestry practice that combines maize, beans, and squash with various tree species. This system helps to reduce soil erosion, improve water retention, and increase crop yields, all while sequestering carbon and supporting biodiversity. The Quesungual system has been recognized as a model for climate-smart agriculture in Central America.

The Taungya System in West Africa

Originating in West Africa, the Taungya system is a form of agroforestry that involves intercropping food crops with tree seedlings. This practice allows farmers to cultivate food crops during the early years of tree growth, providing them with a source of income while the trees mature. The Taungya system has been widely adopted in countries such as Nigeria, Ghana, and Cameroon, where it has contributed to reforestation efforts and improved livelihoods for local communities.

The Dehesa System in Spain

The Dehesa system is a traditional agroforestry practice found in the Iberian Peninsula, characterized by the integration of grazing livestock, such as sheep and cattle, with oak trees. The oak trees provide valuable shade and forage for the animals, while their acorns are a vital ingredient in the production of the world-renowned *Iberian ham*. The Dehesa system demonstrates the potential of silvopasture to support both ecological and economic sustainability.

The Alley Cropping System in the United States

Alley cropping, a form of agroforestry practiced in the United States, involves planting rows of trees or shrubs alongside annual or perennial crops. This system provides numerous benefits, including improved soil health, increased biodiversity, and enhanced crop yields. In the Midwest, for example, farmers have successfully integrated alley cropping with corn and soybean production, demonstrating the compatibility of agroforestry with conventional agricultural practices.

These innovative agroforestry systems showcase the incredible potential of integrating trees and crops to create more sustainable, resilient, and productive landscapes. By learning from these success stories, we can continue to develop and refine agroforestry practices that address the unique challenges and opportunities of different regions around the world.

Silvopasture Success Stories: Livestock and Trees Working Together

In this section, we will delve into some remarkable success stories of silvopasture systems from around the world. These examples demonstrate the potential of integrating livestock and trees harmoniously and sustainably, leading to increased productivity, improved animal welfare, and enhanced environmental benefits.

The Savory Institute's Africa Centre for Holistic Management (ACHM) in Zimbabwe

The ACHM is a shining example of how silvopasture can transform degraded landscapes into thriving ecosystems. Through the implementation of holistic planned grazing, the ACHM has successfully regenerated over 6,500 hectares of land, increasing the carrying capacity for livestock and wildlife. Integrating trees and shrubs into grazing areas has provided shade, improved soil health, and increased biodiversity. This innovative approach has also led to the establishment of a wildlife conservancy, providing a sanctuary for endangered species such as the African wild dog.

The Willowbank Farm in New Zealand

On the North Island of New Zealand, Willowbank Farm has become a model for sustainable silvopasture systems. By integrating livestock, trees, and crops, the farm has achieved increased productivity and resilience against climate change. The farm's innovative approach includes planting high-value timber trees among grazing pastures, providing shelter for livestock, and enhancing soil fertility. This system has improved animal welfare, reduced erosion, and increased carbon sequestration.

The Las Canadas Cooperative in Mexico

The Las Canadas Cooperative is a community-based agroecological project in Veracruz, Mexico that has successfully implemented silvopasture systems to improve the livelihoods of local farmers. By incorporating trees and shrubs into livestock grazing areas, the cooperative has increased forage production, improved soil health, and enhanced biodiversity. The integration of livestock and trees has also provided additional income streams for farmers, such as honey production and ecotourism.

The Montado System in Portugal

The Montado system is a traditional silvopasture practice in Portugal, characterized by the integration of cork and holm oak trees with grazing livestock, primarily sheep and cattle. This ancient system has been recognized for its ability to maintain high levels of biodiversity, preserve cultural heritage, and provide valuable ecosystem services. The Montado system has also proven to be resilient against climate change, as the trees provide shade and help retain soil moisture, ensuring the survival of both livestock and trees during drought.

These success stories demonstrate the immense potential of silvopasture systems in promoting sustainable agriculture and improving the livelihoods of farmers worldwide. By integrating livestock and trees, these systems can enhance productivity, improve animal welfare, and provide valuable environmental benefits. As we continue to face the challenges of climate change and increasing global food demand, adopting silvopasture practices will play a crucial role in shaping the future of sustainable agriculture.

The Role of Technology in Modern Agroforestry and Silvopasture Practices

In today's fast-paced world, technology has become an essential tool in various industries, including agriculture. The integration of technology in agroforestry and silvopasture practices has revolutionized these systems and contributed to their success. This section delves into the role of technology in modern agroforestry and silvopasture practices, highlighting the various innovations that have made these systems more efficient, sustainable, and profitable.

One of the most significant technological advancements in agroforestry and silvopasture is the use of **Geographic Information Systems (GIS) and remote sensing.** These tools enable farmers and researchers to map and analyze land use patterns, monitor vegetation growth, and assess the overall health of the agroforestry or silvopasture system. By providing accurate and real-time data, GIS and remote sensing technologies help farmers make informed decisions about land management, tree planting, and livestock grazing.

Another game-changing innovation in agroforestry and silvopasture is **precision agriculture.** This technology allows farmers to apply the right amount of inputs, such as water, fertilizers, and pesticides, at the right time and place. Precision agriculture techniques, such as drip irrigation and variable rate technology, have significantly reduced water and chemical usage, leading to increased efficiency and reduced environmental impact. Moreover, precision agriculture has also been instrumental in optimizing the productivity of agroforestry and silvopasture systems by enabling farmers to monitor and manage tree and crop growth more effectively.

The use of **drones** or Unmanned Aerial Vehicles (UAVs) has also revolutionized agroforestry and silvopasture practices. Drones equipped with high-resolution cameras and sensors can capture detailed images and data of the land, helping farmers monitor tree health, detect pest infestations, and assess soil conditions. This information is invaluable for making timely interventions and ensuring the long-term success of the agroforestry or silvopasture system.

In addition to these cutting-edge technologies, **mobile applications and software** have become indispensable tools for modern agroforestry and silvopasture practitioners. These digital platforms offer a wealth of information on best practices, market trends, and weather forecasts, enabling farmers to make data-driven decisions and stay connected with the global agricultural community.

Lastly, the rise of social media and online forums has played a crucial role in promoting knowledge exchange and collaboration among agroforestry and silvopasture practitioners worldwide. Farmers can share their experiences through these platforms, learn from each other's successes and challenges, and access valuable resources to enhance their agroforestry or silvopasture systems.

In conclusion, technology has significantly transformed agroforestry and silvopasture practices, making them more efficient, sustainable, and profitable. As we look to the future, it is essential to continue investing in technological innovations and fostering a culture of knowledge sharing to ensure the long-term success of these systems and their role in promoting sustainable agriculture.

Lessons Learned from Agroforestry and Silvopasture Implementations

As we have explored the various case studies of agroforestry and silvopasture systems, reflecting on the lessons learned from these implementations is essential. By understanding the challenges and successes of these systems, we can continue to improve and promote sustainable agriculture practices worldwide.

Customization is key: One of the most significant lessons learned from agroforestry and silvopasture implementations is that there is no one-size-fits-all solution. Each system must be tailored to the specific needs of the local environment, culture, and economy. This requires a deep understanding of the local ecology, climate, and social dynamics to create a productive and sustainable system.

Collaboration and knowledge sharing: The success of agroforestry and silvopasture systems often depends on the collaboration between farmers, researchers, and policymakers. By working together and sharing knowledge, we can develop innovative solutions that address the unique challenges faced by each community. This collaborative approach also helps to build trust and foster a sense of ownership among stakeholders, which is crucial for the long-term success of these systems.

Adaptability and resilience: Agroforestry and silvopasture systems must be adaptable to changing environmental conditions, such as climate change and shifting market demands. This requires ongoing monitoring, evaluation, and adaptation of management practices to ensure the system remains productive and sustainable. By building resilience into these systems, we can help to buffer against the impacts of climate change and other external pressures.

Integration of traditional and modern practices: Many successful agroforestry and silvopasture systems draw on a combination of traditional knowledge and modern scientific research. By integrating these two sources of knowledge, we can create systems that are both culturally appropriate and scientifically robust. This approach also helps to preserve and promote traditional agricultural practices, which are often more sustainable and better suited to local conditions than conventional farming methods.

Economic viability: For agroforestry and silvopasture systems to be widely adopted, they must be economically viable for farmers. This requires the development of markets for the diverse range of products that these systems can produce, such as timber, non-timber forest products, and livestock products. By creating value-added opportunities and supporting the development of local and regional markets, we can help to ensure the long-term economic sustainability of these systems.

Education and capacity building: The success of agroforestry and silvopasture systems depends on the skills and knowledge of the farmers who manage them. By providing training, technical assistance, and ongoing support, we can build the capacity of farmers to implement and manage these systems effectively. This includes practical skills, such as tree planting and livestock management, and broader knowledge about ecology, climate change, and market dynamics.

In conclusion, the lessons learned from agroforestry and silvopasture implementations highlight the importance of customization, collaboration, adaptability, integration of traditional and modern practices, economic viability, and education. By applying these lessons, we can continue to develop and promote sustainable agriculture practices that benefit both people and the planet.

The Future of Sustainable Agriculture through Agroforestry and Silvopasture

As we have explored throughout this chapter, agroforestry and silvopasture systems have proven to be highly effective and sustainable agricultural practices. By integrating trees and livestock into agricultural systems, these practices have successfully addressed various environmental, social, and economic challenges faced by farmers and communities around the world. As we look towards the future, it is clear that agroforestry and silvopasture will continue to play a crucial role in shaping sustainable agriculture and ensuring food security for a growing global population.

One of the key takeaways from the case studies presented in this chapter is the incredible **diversity and adaptability of agroforestry and silvopasture systems**. From small-scale family farms to large commercial operations, these practices have been successfully implemented across various landscapes, climates, and cultures. This adaptability is essential for the future of sustainable agriculture, as it allows farmers to tailor their agroforestry and silvopasture systems to their specific needs and local conditions.

Moreover, agroforestry and silvopasture systems have demonstrated their ability to provide multiple benefits, including increased biodiversity, improved soil health, enhanced carbon sequestration, and reduced erosion. These environmental benefits contribute to the overall health and resilience of agricultural landscapes and help mitigate the impacts of climate change. As the global community continues to grapple with the challenges posed by climate change, adopting agroforestry and silvopasture practices will become increasingly important in building a more sustainable and resilient food system.

In addition to their environmental benefits, agroforestry and silvopasture systems offer significant **social and economic advantages** for farmers and rural communities. By diversifying agricultural production, these systems can help farmers increase their income, reduce their vulnerability to market fluctuations, and improve their livelihoods. Furthermore, integrating trees and livestock into agricultural systems can create new opportunities for rural employment and contribute to the revitalization of rural communities.

As we have seen in the case studies, technology also plays a vital role in the successful implementation of agroforestry and silvopasture systems. From precision agriculture tools to innovative livestock management techniques, technology has the potential to revolutionize these practices and make them more accessible, efficient, and profitable for farmers. As technology continues to advance, it will be essential for researchers, policymakers, and practitioners to work together to ensure that these innovations are integrated into agroforestry and silvopasture systems in a way that supports their long-term sustainability and success.

In conclusion, the future of sustainable agriculture lies in the continued development and adoption of agroforestry and silvopasture practices. By learning from the successes and challenges of the case studies presented in this chapter, we can continue to refine and expand these systems to meet the diverse needs of farmers, communities, and ecosystems around the world. As we move forward, we must continue to invest in research, education, and policy support for agroforestry and silvopasture, ensuring that these practices remain at the forefront of our efforts to build a more sustainable and resilient global food system.

Chapter Summary

- Agroforestry and silvopasture systems integrate trees, shrubs, and livestock into agricultural landscapes, offering numerous benefits such as increased biodiversity, improved soil health, and enhanced carbon sequestration.
- Customization is vital for successful agroforestry and silvopasture implementations, as each system must be tailored to the specific needs of the local environment, culture, and economy.
- Collaboration and knowledge sharing among farmers, researchers, and policymakers are essential for developing innovative solutions and the long-term success of these systems.
- Adaptability and resilience are crucial for agroforestry and silvopasture systems to withstand changing environmental conditions, such as climate change and shifting market demands.
- Integration of traditional and modern practices, including technology, can create culturally appropriate and scientifically robust systems, preserving traditional agricultural practices while enhancing efficiency.
- Economic viability is necessary for the widespread adoption of agroforestry and silvopasture systems, requiring the development of markets for diverse products and value-added opportunities.
- Education and capacity building are vital for farmers to effectively implement and manage agroforestry and silvopasture systems, including practical skills and broader

knowledge about ecology, climate change, and market dynamics.

- The future of sustainable agriculture lies in the continued development and adoption of agroforestry and silvopasture practices, with technology playing a vital role in revolutionizing these systems and making them more accessible, efficient, and profitable for farmers.

10

FUTURE TRENDS AND INNOVATIONS IN AGROFORESTRY AND SILVOPASTURE

A futuristic scenario where agroforestry and silvopasture systems have overcome their implementation challenges and have become the norm in sustainable agriculture.

As we stand on the precipice of a new era in agriculture and land management, we must recognize and embrace the immense potential of agroforestry and silvopasture. These innovative systems, which integrate trees and shrubs with crops and livestock, have the power to transform our landscapes, enhance biodiversity, and promote sustainable food production. In this chapter, we will explore the future trends and innovations in agroforestry and silvopasture, delving into the exciting possibilities that lie ahead.

Agroforestry and silvopasture are not new concepts; they have been practiced for centuries by indigenous communities and small-scale farmers across the globe. However, in recent years, these systems have gained renewed attention and interest from researchers, policymakers, and farmers alike. This resurgence is driven by a growing awareness of the pressing need to address the challenges of climate change, food security, and environmental degradation.

The beauty of agroforestry and silvopasture lies in their **ability to provide multiple benefits simultaneously**. Combining trees with crops and livestock can increase agricultural productivity, improve soil health, conserve water, enhance wildlife habitat, and sequester carbon. Moreover, they offer a means to diversify farm income, reduce risk, and build resilience in the face of a changing climate.

In this chapter, we will delve into the future trends and innovations that are shaping the world of agroforestry and silvopasture. We will explore the technological advancements that are revolutionizing the way these systems are managed and the role of policy and incentives in driving their adoption. We will also examine the importance of integrating indigenous knowledge and practices into modern agroforestry and silvopasture systems, recognizing the valuable insights that these traditional land management systems can offer.

As we journey through this chapter, we invite you to envision a future where agroforestry and silvopasture are widely adopted, playing a central role in creating sustainable and resilient landscapes. By embracing the potential of these innovative systems, we can work together to address the pressing challenges of our time and chart a course toward a more sustainable and prosperous future for all.

Technological Advancements in Agroforestry and Silvopasture Management

The integration of technology has become an essential aspect of modern agriculture, and agroforestry and silvopasture are no exception. Technological advancements have the potential to revolutionize the way we manage these systems, making them more efficient, productive, and sustainable. This section will explore some of the most promising innovations in agroforestry and silvopasture management.

One of the most significant advancements in recent years is the use of **remote sensing and Geographic Information Systems (GIS)** in agroforestry and silvopasture management. These tools allow for the collection and analysis of spatial data, enabling land managers to make informed decisions about the design and management of their systems. For example, remote sensing can be used to monitor tree growth, assess the health of vegetation, and identify areas that require intervention. GIS, on the other hand, can be used to map and analyze the distribution of different tree species, helping land managers optimize the spatial arrangement of their agroforestry systems.

Another promising innovation is the use of **precision agriculture** techniques in agroforestry and silvopasture management. Precision agriculture involves the use of technology to collect and analyze data on various aspects of farming, such as soil fertility, plant health, and weather conditions. This information can then be used to make more informed decisions about the management of agroforestry and silvopasture systems, such as when to plant, fertilize, or harvest. For example, precision agriculture technologies like soil sensors can help land managers monitor soil moisture and nutrient levels, allowing them to apply water and fertilizer more efficiently and reduce waste.

Drones, or unmanned aerial vehicles (UAVs), are another technological innovation that has the potential to transform agroforestry and silvopasture management. Drones can be equipped with cameras and sensors to collect high-resolution images and data on tree growth, vegetation health, and animal movements. This information can be used to monitor the health of agroforestry and silvopasture systems, identify areas that require intervention, and assess the effectiveness of management practices. Additionally, drones can be used to perform tasks such as planting seeds, spraying pesticides, and monitoring livestock, making the management of these systems more efficient and less labor-intensive.

Finally, the use of **digital platforms and mobile applications** is becoming increasingly important in agroforestry and silvopasture management. These tools can help land managers access information on best practices, connect with other practitioners, and share their experiences and knowledge. For example, mobile applications can guide tree species selection, assist with the design of agroforestry systems, and offer advice on the management of silvopasture systems. Digital platforms can also facilitate the exchange of information and resources between researchers, policymakers, and practitioners, fostering collaboration and innovation in the field of agroforestry and silvopasture.

In conclusion, technological advancements are playing a crucial role in the evolution of agroforestry and silvopasture management. By embracing these innovations, land managers can optimize the design and management of their systems, making them more productive, sustainable, and resilient in the face of future challenges.

Climate Change Adaptation and Mitigation through Agroforestry and Silvopasture

As the world grapples with the ever-growing challenge of climate change, it is becoming increasingly evident that agroforestry and silvopasture systems have a crucial role to play in both adaptation and mitigation efforts. These integrated land-use practices offer a sustainable approach to agriculture and livestock management and provide various environmental benefits that can help reduce greenhouse gas emissions and bolster the resilience of ecosystems and communities.

One of the most significant ways in which agroforestry and silvopasture contribute to climate change mitigation is through **carbon sequestration**. Trees and other perennial plants in these systems capture and store atmospheric carbon dioxide in their biomass and in the soil, effectively acting as carbon sinks. Research has shown that well-managed agroforestry systems can sequester up to four times more carbon than conventional agricultural systems, making them a powerful tool in the fight against climate change.

In addition to their carbon sequestration potential, agroforestry and silvopasture systems can also help to **reduce greenhouse gas emissions** from agriculture. By incorporating trees and shrubs into agricultural landscapes, these systems can improve nutrient cycling and reduce the need for synthetic fertilizers, which are a major source of nitrous oxide emissions. Moreover, the shade trees provide in silvopasture systems can help reduce methane emissions from livestock by lowering their body temperature and decreasing enteric fermentation.

Agroforestry and silvopasture systems also offer a range of adaptation benefits, helping farmers and communities to build resilience in the face of climate change. For instance, the presence of trees can help to buffer crops and livestock from extreme weather events, such as droughts, floods, and heatwaves. Trees can also improve water infiltration and retention in the soil, reducing the impacts of water scarcity and ensuring a more stable water supply for agriculture.

Furthermore, the increased biodiversity associated with agroforestry and silvopasture systems can enhance the resilience of ecosystems and agricultural production. A diverse mix of species can provide a range of ecological services, such as pest control, pollination, and soil fertility, reducing the vulnerability of farms to climate-induced disturbances. This diversity can also help to spread the risk of crop failure, ensuring greater food security for communities in the face of climate change.

In conclusion, agroforestry and silvopasture systems offer a promising pathway for climate change adaptation and mitigation in the agricultural sector. By harnessing the power of trees and integrated land-use practices, these systems can help to reduce greenhouse gas emissions, sequester carbon, and build resilience in ecosystems and communities. Therefore, as we look to the future, we must continue to explore and promote the potential of agroforestry and silvopasture as key components of a sustainable and climate-resilient agricultural landscape.

The Role of Policy and Incentives in Shaping the Future of Agroforestry and Silvopasture

As we look toward the future of agroforestry and silvopasture, we must recognize the critical role that policy and incentives play in shaping these systems' growth and development. Governments, international organizations, and private institutions can all contribute to fostering a supportive environment for the widespread adoption of agroforestry and silvopasture practices. We can accelerate the transition towards more sustainable and resilient agricultural systems by implementing well-designed policies and incentives.

One of the most significant ways that policy can influence the future of agroforestry and silvopasture is by providing **financial incentives** to farmers and landowners. These incentives can take various forms, such as grants, subsidies, tax breaks, or low-interest loans, and can help offset the initial costs of establishing agroforestry or silvopasture systems. By reducing the financial barriers to entry, these incentives can encourage more farmers to adopt these practices and help them see the long-term benefits of integrating trees and livestock into their agricultural systems.

In addition to financial incentives, policy can also play a vital role in promoting research and development in agroforestry and silvopasture. By investing in scientific research and technological innovation, governments and organizations can help advance our understanding of these systems and develop new tools and techniques to optimize their management. This research can also help identify the most effective agroforestry and silvopasture practices for different regions and climates, ensuring these systems are tailored to local conditions and needs.

Another crucial aspect of policy is the development of clear **guidelines and regulations** for agroforestry and silvopasture practices. By establishing standards for designing, implementing, and managing these systems, governments can ensure that they are environmentally sustainable and socially responsible. These guidelines can also help farmers and landowners navigate the complexities of agroforestry and silvopasture, providing them with the information and resources they need to succeed.

Finally, policy can play a vital role in fostering collaboration and knowledge-sharing among stakeholders in the agroforestry and silvopasture sectors. By creating **platforms for dialogue and cooperation**, governments and organizations can facilitate the exchange of ideas, experiences, and best practices among farmers, researchers, and policymakers. This collaborative approach can help build a strong and supportive community around agroforestry and silvopasture, ensuring that these systems continue to evolve and adapt to the changing needs of our world.

In conclusion, the future of agroforestry and silvopasture will be significantly influenced by the policies and incentives that are put in place to support their growth and development. By recognizing the potential of these systems and investing in their success, we can create a more sustainable and resilient future for agriculture and the environment.

Integrating Indigenous Knowledge and Practices into Modern Agroforestry and Silvopasture Systems

The importance of indigenous knowledge and practices in agroforestry and silvopasture systems cannot be overstated. As custodians of the land for generations, indigenous communities have developed a wealth of knowledge and expertise in managing their natural resources sustainably. This section delves into the significance of integrating indigenous knowledge and practices into modern agroforestry and silvopasture systems, highlighting the benefits and challenges of doing so.

Indigenous communities have long understood the intricate relationships between plants, animals, and their environment. They have developed systems that provide food, fodder, and fuel, maintain the ecological balance, and enhance biodiversity. These traditional agroforestry and silvopasture systems are often characterized by a high degree of species diversity, multiple layers of vegetation, and integration of crops, trees, and livestock.

Integrating indigenous knowledge and practices into modern agroforestry and silvopasture systems can provide **numerous benefits.** Firstly, it can help improve the resilience and adaptability of these systems in the face of climate change. Indigenous communities have long adapted their practices to cope with changing environmental conditions. Therefore, their knowledge can provide valuable insights into how to design and manage agroforestry and silvopasture systems that can withstand the impacts of climate change.

Secondly, incorporating indigenous knowledge and practices can enhance the ecological sustainability of agroforestry and silvopasture systems. Traditional systems often prioritize the conservation of soil, water, and biodiversity, and integrating these principles into modern systems can help ensure that they are managed to minimize negative environmental impacts.

Thirdly, integrating indigenous knowledge and practices can contribute to the social and cultural sustainability of agroforestry and silvopasture systems. By recognizing and valuing the knowledge and expertise of indigenous communities, we can foster a sense of ownership and pride in these systems, which can, in turn, encourage their continued stewardship and sustainable management.

Despite the clear benefits of integrating indigenous knowledge and practices into modern agroforestry and silvopasture systems, there are also challenges to overcome. One key challenge is the need to **bridge the gap** between traditional and scientific knowledge. This requires fostering dialogue and collaboration between indigenous communities, researchers, and practitioners to ensure that indigenous knowledge is respected and incorporated in a way that complements and enhances scientific understanding.

Another challenge is the need to **protect the intellectual property rights** of indigenous communities. As their knowledge and practices are increasingly recognized for their value in sustainable land management, there is a risk of exploitation and appropriation. Therefore, it is essential to develop mechanisms that ensure the fair and equitable sharing of benefits arising from the use of indigenous knowledge and practices.

In conclusion, integrating indigenous knowledge and practices into modern agroforestry and silvopasture systems holds great promise for enhancing their sustainability, resilience, and adaptability. By fostering collaboration and dialogue between indigenous communities, researchers, and practitioners, we can ensure that this valuable knowledge is harnessed to create agroforestry and silvopasture systems that are not only productive but also ecologically, socially, and culturally sustainable.

Envisioning a Sustainable and Resilient Future for Agroforestry and Silvopasture

As we reach the end of our exploration into the world of agroforestry and silvopasture, it is essential to envision a future where these practices become the norm rather than the exception. The potential of agroforestry and silvopasture to address some of the most pressing challenges of our time, such as climate change, food security, and biodiversity loss, is immense. By integrating trees, crops, and livestock, we can create more resilient, diverse, and sustainable landscapes that benefit both people and the planet.

Technological advancements will undoubtedly play a crucial role in the widespread adoption of agroforestry and silvopasture practices. From precision agriculture tools that optimize resource use to innovative tree-crop-livestock combinations that maximize productivity, technology can revolutionize these systems. As we continue to develop and refine these tools, it is crucial to ensure they are accessible and affordable to farmers worldwide, particularly those in developing countries who stand to benefit the most from these practices.

Climate change adaptation and mitigation will also be central to the future of agroforestry and silvopasture. As the impacts of climate change become increasingly apparent, it is essential to recognize the role that these practices can play in building resilience and sequestering carbon. By incorporating trees and other perennial vegetation into agricultural systems, we can create landscapes that are better equipped to withstand the challenges of a changing climate while also contributing to global efforts to reduce greenhouse gas emissions.

Policy and incentives will be instrumental in shaping the future of agroforestry and silvopasture. Governments, international organizations, and the private sector must work together to create an enabling environment that supports the adoption and scaling up of these practices. This includes providing financial incentives, technical assistance, and favorable policy frameworks that recognize the multiple benefits of agroforestry and silvopasture and promote their integration into mainstream agricultural systems.

Finally, the future of agroforestry and silvopasture must be grounded in the **integration of indigenous knowledge and practices.** Indigenous communities have been practicing agroforestry and silvopasture for centuries, and their wisdom and experience can provide valuable insights into the design and management of these systems. By respecting and incorporating this knowledge, we can create more resilient and sustainable agroforestry and silvopasture systems that are better adapted to local conditions and cultural contexts.

In conclusion, the future of agroforestry and silvopasture is bright and filled with opportunities to create more sustainable, resilient, and productive landscapes. By embracing the potential of these practices and working together to overcome the challenges that lie ahead, we can ensure a future where agroforestry and silvopasture play a central role in addressing some of the most pressing issues of our time, ultimately leading to a healthier planet and a more secure future for all.

Chapter Summary

- Agroforestry and silvopasture systems, which integrate trees and shrubs with crops and livestock, have the potential to transform landscapes, enhance biodiversity, and promote sustainable food production.
- Technological advancements, such as remote sensing, GIS, precision agriculture, drones, and digital platforms, are revolutionizing the management of agroforestry and silvopasture systems, making them more efficient and sustainable.
- Agroforestry and silvopasture systems can play a crucial role in climate change adaptation and mitigation by sequestering carbon, reducing greenhouse gas emissions, and enhancing the resilience of ecosystems and communities.
- Policy and incentives are essential in shaping the future of agroforestry and silvopasture, with governments, international organizations, and private institutions all contributing to fostering a supportive environment for their widespread adoption.
- Integrating indigenous knowledge and practices into modern agroforestry and silvopasture systems can enhance their sustainability, resilience, and adaptability while also respecting and valuing the expertise of indigenous communities.
- Challenges in integrating indigenous knowledge and practices include bridging the gap between traditional and scientific knowledge, fostering collaboration between stakeholders, and protecting the intellectual property rights of indigenous communities.
- The future of agroforestry and silvopasture depends on the widespread adoption of these practices, supported by technological advancements, climate change adaptation and

mitigation efforts, favorable policies and incentives, and the integration of indigenous knowledge.

- By embracing the potential of agroforestry and silvopasture systems and working together to overcome challenges, we can create more sustainable, resilient, and productive landscapes, ultimately leading to a healthier planet and a more secure future for all.

THE PATH FORWARD FOR AGROFORESTRY AND SILVOPASTURE

As we reach the conclusion of this enlightening exploration into the world of agroforestry and silvopasture, it is essential to take a moment to reflect on the incredible journey we have embarked upon together. Through the pages of this book, we have delved deep into the history, principles, and practices of these two sustainable land management systems, uncovering the immense potential they hold for transforming our relationship with the environment and ensuring a prosperous future for generations to come.

Agroforestry and silvopasture have emerged as powerful tools in the fight against climate change, land degradation, and biodiversity loss. By integrating trees, crops, and livestock, these systems not only enhance the productivity and resilience of agricultural landscapes but also provide a wide array of ecosystem services, such as carbon sequestration, soil conservation, and habitat creation. Moreover, they offer a unique opportunity to bridge the gap between traditional agricultural practices and modern ecological knowledge, fostering a more harmonious and sustainable coexistence between humans and nature.

However, despite their numerous benefits and growing recognition, agroforestry and silvopasture still need to overcome significant challenges and barriers to widespread adoption. From entrenched mindsets and lack of awareness to inadequate policy support and financial incentives, there is much work to be done to unlock the full potential of these systems and ensure their rightful place in the global agricultural landscape.

In this final chapter, we will explore the path forward for agroforestry and silvopasture, outlining the key steps and strategies needed to overcome these challenges and catalyze a paradigm shift in the way we manage our land and resources. We will discuss the importance of strengthening policy frameworks and incentives, enhancing research, education, and extension efforts, and fostering collaboration and community engagement. Lastly, we will share some final thoughts on the future of agroforestry and silvopasture, envisioning a world where these systems are embraced as cornerstones of a sustainable and resilient global food system.

So, let us embark on this final leg of our journey together, with renewed hope and determination, as we chart a course toward a brighter, greener, and more equitable future for all.

Overcoming Challenges and Barriers to Adoption

As we reflect on the journey of agroforestry and silvopasture, it is crucial to acknowledge the challenges and barriers that have hindered their widespread adoption. To pave the way for a sustainable and resilient future, we must address these obstacles head-on and develop strategies to overcome them.

One of the primary challenges farmers and landowners face is the **need for knowledge and understanding** of agroforestry and silvopasture systems. This can be attributed to the limited availability of resources and training programs tailored to these practices. To overcome this barrier, we must develop comprehensive educational materials and training programs that cater to the unique needs of farmers interested in adopting agroforestry and silvopasture practices. By equipping them with the necessary skills and knowledge, we can empower them to make informed decisions and successfully implement these systems on their land.

Another significant barrier to adoption is the perception of **high initial costs** and the time it takes to see returns on investment. While it is true that establishing agroforestry and silvopasture systems may require a higher upfront investment compared to conventional agricultural practices, it is essential to highlight the long-term economic, social, and environmental benefits that these systems offer. By emphasizing the potential for increased productivity, improved soil health, enhanced biodiversity, and reduced vulnerability to climate change, we can encourage farmers and landowners to view these practices as a worthwhile investment in the future of their land and livelihoods.

The lack of supportive policy frameworks and incentives also poses a challenge to the adoption of agroforestry and silvopasture. To address this issue, we must advocate for the development and implementation of policies and programs that specifically target and promote these practices. This can include financial incentives, such as grants, low-interest loans, and tax breaks, as well as technical assistance and resources to help farmers and landowners navigate the transition to agroforestry and silvopasture systems.

Lastly, the absence of strong networks and collaboration among stakeholders has hindered the growth and development of agroforestry and silvopasture. By fostering a sense of community and collaboration among farmers, researchers, policymakers, and other stakeholders, we can create a supportive environment that enables the exchange of knowledge, resources, and experiences. This collaborative approach will not only help overcome the challenges faced by individual farmers but also contribute to the broader advancement of agroforestry and silvopasture as sustainable and resilient land management practices.

In conclusion, overcoming the challenges and barriers to the adoption of agroforestry and silvopasture requires a multifaceted approach that addresses the needs of farmers and landowners, strengthens policy frameworks and incentives, enhances research and education efforts, and fosters collaboration and community engagement. By tackling these obstacles head-on, we can pave the way for a more sustainable and resilient future for agroforestry and silvopasture.

Strengthening Policy Frameworks and Incentives

As we reflect on the journey of agroforestry and silvopasture, we must recognize the critical role that policy frameworks and incentives play in promoting their adoption and success. By strengthening these mechanisms, we can create a more supportive environment for farmers, landowners, and other stakeholders to embrace these sustainable land management practices. In this section, we will discuss the importance of policy frameworks and incentives, and outline some strategies to bolster them for the benefit of agroforestry and silvopasture.

Policy frameworks are the foundation upon which agroforestry and silvopasture initiatives can be built. They provide the necessary guidelines, regulations, and support systems to ensure these practices are implemented effectively and sustainably. To strengthen policy frameworks, governments and policymakers must prioritize the development and implementation of comprehensive policies that specifically address the unique needs and challenges of agroforestry and silvopasture. This includes integrating these practices into existing agricultural, forestry, and environmental policies, as well as creating new policies that specifically cater to their promotion and growth.

One key aspect of strengthening policy frameworks is the provision of **financial incentives** for farmers and landowners who adopt agroforestry and silvopasture practices. These incentives can take various forms, such as grants, subsidies, tax breaks, and low-interest loans. By offering financial support, governments can offset the initial costs of establishing these systems and make them more attractive to potential adopters. Additionally, incentives can be designed to reward long-term commitment to agroforestry and silvopasture, ensuring that these practices continue to provide environmental, social, and economic benefits over time.

Another crucial component of policy frameworks is the establishment of **clear and enforceable regulations** that protect the integrity of agroforestry and silvopasture systems. This includes setting standards for managing these systems, monitoring their performance, and enforcing penalties for non-compliance. By doing so, governments can ensure that agroforestry and silvopasture practices are carried out responsibly and sustainably, thereby maximizing their potential benefits.

In addition to strengthening policy frameworks, it is also essential to enhance education and extension efforts to promote agroforestry and silvopasture. This includes providing training and technical assistance to farmers and landowners and raising awareness about the benefits of these practices among the general public. By doing so, we can foster a greater understanding of agroforestry and silvopasture and encourage more widespread adoption of these practices.

In conclusion, strengthening policy frameworks and incentives is a vital step in the path forward for agroforestry and silvopasture. By creating a supportive environment for these practices, we can help ensure their continued growth and success, ultimately contributing to a more sustainable and resilient future for our agricultural and forestry systems.

Enhancing Research, Education, and Extension Efforts

As we continue to explore the potential of agroforestry and silvopasture in creating a sustainable and resilient future, we must invest in enhancing research, education, and extension efforts. This will not only help in overcoming the existing challenges and barriers but also pave the way for innovation and widespread adoption of these practices.

First and foremost, we need to prioritize research that focuses on the development of new agroforestry and silvopasture systems, as well as the improvement of existing ones. This includes studying the ecological, social, and economic aspects of these systems and understanding how they can be optimized to provide maximum benefits to farmers, communities, and the environment. By investing in cutting-edge research, we can identify best practices, develop new technologies, and create models that can be replicated and scaled up across different regions and contexts.

In tandem with research, it is essential to **strengthen education**

and training programs related to agroforestry and silvopasture. This includes incorporating these topics into the curricula of agricultural and environmental science programs, as well as offering specialized courses and workshops for farmers, land managers, and other stakeholders. By fostering a deeper understanding of the principles, practices, and benefits of agroforestry and silvopasture, we can empower the next generation of professionals and practitioners to become champions of these systems.

Extension efforts play a critical role in bridging the gap between research and practice, ensuring that the knowledge and innovations generated through research are translated into tangible, on-the-ground actions. To enhance extension efforts, we need to invest in building the capacity of extension agents and other intermediaries, equipping them with the tools, resources, and skills needed to effectively promote and support the adoption of agroforestry and silvopasture practices. This includes developing user-friendly guides, manuals, and decision-support tools, as well as leveraging digital platforms and technologies to reach a wider audience.

Moreover, it is essential to recognize that knowledge exchange is a two-way street, and that farmers and land managers have valuable insights and experiences to share. By fostering a culture of participatory research and co-learning, we can ensure that the development and dissemination of agroforestry and silvopasture practices are grounded in the realities and needs of those directly involved in their implementation.

In conclusion, enhancing research, education, and extension efforts is vital to realizing the full potential of agroforestry and silvopasture as sustainable land management practices. By investing in these areas, we can generate the knowledge, skills, and innovations needed to drive the widespread adoption of these systems, ultimately contributing to a more sustainable and resilient future for our planet.

Fostering Collaboration and Community Engagement

As we approach the culmination of our exploration into the realms of agroforestry and silvopasture, it is essential to recognize the importance of collaboration and community engagement in ensuring the success and widespread adoption of these practices. The sustainable and resilient future we envision for agroforestry and silvopasture can only be achieved through the collective efforts of various stakeholders, including farmers, researchers, policymakers, and local communities.

The first step in fostering collaboration is to create platforms and opportunities for **dialogue and knowledge exchange** among these diverse stakeholders. By facilitating open communication, we can bridge gaps in understanding, identify shared goals, and develop strategies to overcome the challenges and barriers to adoption discussed earlier in this chapter. Workshops, conferences, and field days can serve as valuable venues for such interactions, allowing stakeholders to learn from each other's experiences and insights.

In addition to dialogue, collaboration can be further strengthened by establishing partnerships and networks that bring together stakeholders with complementary skills and resources. For example, farmers can collaborate with researchers to conduct on-farm trials and demonstrations, generating valuable data and practical knowledge that can inform the development of improved agroforestry and silvopasture systems. Similarly, partnerships between policymakers and local communities can ensure that policy frameworks and incentives are tailored to the specific needs and contexts of different regions.

Community engagement is another critical aspect of fostering collaboration, as it empowers local stakeholders to take ownership of agroforestry and silvopasture initiatives and contribute to their success. By involving communities in the planning, implementing, and monitoring of these practices, we can ensure that their perspectives, knowledge, and priorities are taken into account, leading to more sustainable and context-appropriate solutions. Moreover, community engagement can help build trust and social capital, which are essential for the long-term viability of agroforestry and silvopasture systems.

One effective approach to community engagement is the **establishment of local agroforestry and silvopasture groups or associations.** These groups can serve as platforms for peer-to-peer learning, mentorship, and support, enabling farmers and other community members to share their experiences, challenges, and successes with one another. Furthermore, these groups can play a crucial role in advocating for agroforestry and silvopasture at the local, regional, and national levels, raising awareness of the benefits of these practices and influencing policy decisions.

In conclusion, fostering collaboration and community engagement is a vital component of the path forward for agroforestry and silvopasture. By working together, leveraging our collective strengths, and engaging local communities, we can accelerate the adoption of these practices and move closer to realizing our vision of a sustainable and resilient future for agriculture and the environment.

Final Thoughts: Envisioning a Sustainable and Resilient Future for Agroforestry and Silvopasture

As we reach the end of this enlightening journey through the world of agroforestry and silvopasture, it is essential to take a moment to reflect on the potential of these practices in shaping a sustainable and resilient future for our planet. The integration of trees, crops, and livestock not only holds the promise of increased productivity and profitability for farmers but also offers a myriad of environmental and social benefits that cannot be overlooked.

In the face of climate change, dwindling natural resources, and growing global food demand, adopting innovative and sustainable land management practices is more important than ever. Agroforestry and silvopasture have the potential to be at the forefront of this change, providing solutions to some of the most pressing challenges of our time. By sequestering carbon, improving soil health, conserving biodiversity, and enhancing water quality, these systems can play a vital role in mitigating the impacts of climate change and preserving the environment for future generations.

However, realizing this potential requires overcoming the various challenges and barriers to adoption that we have discussed throughout this book. It is crucial to strengthen policy frameworks and incentives that support the widespread implementation of agroforestry and silvopasture practices. Governments, NGOs, and private sector stakeholders must work together to create an enabling environment that encourages farmers to embrace these systems and invest in their long-term success.

Moreover, enhancing research, education, and extension efforts is vital to unlocking the full potential of agroforestry and silvopasture. By deepening our understanding of these practices and their benefits, we can develop more effective strategies for their implementation and management. At the same time, fostering collaboration and community engagement is essential to ensure farmers have access to the knowledge, resources, and support they need to succeed in their agroforestry and silvopasture endeavors.

In conclusion, the path forward for agroforestry and silvopasture is one of immense opportunity and promise. By harnessing the power of these systems, we can create a more sustainable and resilient future for agriculture and the environment. It is our **collective responsibility** to seize this moment and work together to make this vision a reality. As we embark on this journey, let us be guided by the principles of innovation, collaboration, and sustainability, and let us never lose sight of the ultimate goal: a thriving planet for all.

ABOUT THE AUTHOR

Michael Barton is an expert in sustainability and regenerative agriculture with over a decade of experience. He holds a degree in Environmental Studies and has received advanced training in the principles and practices of regenerative agriculture.

Throughout his career, Michael has worked with farmers, policymakers, and environmentalists to promote sustainable farming practices and advocate for the adoption of regenerative agriculture. He has collaborated with organizations worldwide to advance sustainable agriculture and food systems.

In his free time, Michael enjoys hiking and spending time with his family.

Made in the USA
Monee, IL
04 November 2026

67383023R00167